THE SPACE RACE

THE SPACE RACE

FOREWORD

To celebrate the turn of the century and the new millennium, **THE EVENTFUL CENTURY** series presents the vast panorama of the last hundred years—a century which has witnessed the transition from horse-drawn transport to space travel, and from the first telephones to the information superhighway.

THE EVENTFUL CENTURY chronicles epoch-making events like the outbreak of the two world wars, the Russian Revolution and the rise and fall of communism. But major events are only part of this glittering kaleidoscope. It also describes the everyday background—the way people lived, how they worked, what they ate and drank, how much they earned, the way they spent their leisure time, the books they read, and the crimes, scandals and unsolved mysteries that set them talking. Here are fads and crazes like the Hula-Hoop and Rubik's Cube . . . fashions like the New Look and the miniskirt . . . breakthroughs in entertainment, such as the birth of the movies . . . medical milestones such as the discovery of penicillin . . . and marvels of modern architecture and engineering.

THE SPACE RACE tells the remarkable story of one of the most rapid scientific and technological developments in the history of the world. At the dawn of the twentieth century, no one had yet flown in a powered aircraft, and the fastest cars could manage only 25 mph. The notion of traveling into space was the preserve of science-fiction writers, dreamers and cranks. Yet only 57 years later, the first living creature from this planet—a dog named Laika—was orbiting the Earth in a primitive capsule. Then, in 1961, Yuri Gagarin became the first human being in space, and the race was on in earnest for the rest of the decade. Space development in the 1960s culminated in the awesome moment when Neil Armstrong stepped onto the surface of the Moon, having been transported there by a vehicle of then unprecedented sophistication. By the end of the century, space probes and super-powered telescopes were revealing not just the secrets of our own Solar System, but of the whole Universe. This volume follows the personalities who dared to dream the impossible, and traces the triumphs and disasters that led to the Moon, to Mars and now beyond.

THE SPACE RACE

The Reader's Digest Association, Inc.
Pleasantville, New York/Montreal

THE SPACE RACE
Edited and designed by Toucan Books Limited
Written by John Man
Edited by Helen Douglas-Cooper and
David Scott-Macnab
Designed by Bradbury and Williams
Picture research by Robert Sackville West

FOR THE AMERICAN EDITION
Produced by The Reference Works, Inc.
Director Harold Rabinowitz
Editors Geoffrey Upton, Lorraine Martindale
Production Antler DesignWorks
Director Bob Antler

FOR READER'S DIGEST
Group Editorial Director Fred DuBose
Senior Editor Susan Randol
Senior Designers Carol Nehring, Judith Carmel
Production Technology Manager Douglas A. Croll
Art Production Coordinator Jennifer R. Tokarski

READER'S DIGEST ILLUSTRATED REFERENCE BOOKS
Editor-in-Chief Christopher Cavanaugh
Art Director Joan Mazzeo

Library of Congress
Cataloging in Publication Data:
The space race
    p. cm. — (The eventful 20th century)
  ISBN 0-7621-0287-X
    1. Space race—History. 2. Astronautics—History.
    3. Technological innovations—History—20th
    century. I. Reader's Digest Association. II. Series.

  TL788.5 .S637 2000
  629.41'09—dc21
                                    00-028089

FRONT COVER
From Top: Buzz Aldrin on the Moon; Mariner 2;
Senator John Glenn.

BACK COVER
From Top: Space Shuttle lift-off; Skylab mission
badge; John Glenn with President Kennedy and
Vice-President Johnson.

Page 3 (from left to right): Rudolf Nebel; Apollo
badge; Lunar Orbiter 1; John Young and Robert
Crippen with a model of Space Shuttle *Columbia*.

Background pictures:
Page 11: Tsiolkovsky's calculations.
Page 29: Major White makes a space walk, 1966.
Page 67: View of the Moon from Apollo 8.
Page 99: Jupiter with Io and Europa.
Page 119: On board Skylab 4.

Address any comments about The Space Race to
Reader's Digest, Editor-in-Chief, U.S. Illustrated
Reference Books,
Reader's Digest Road, Pleasantville, NY 10570

To order additional copies of The Space Race, call
1-800-846-2100

You can also visit us on the World Wide Web at:
www.readersdigest.com

CONTENTS

THE SPIRIT OF DISCOVERY

WITHIN THE SPAN OF A SINGLE HUMAN LIFETIME, SPACE EXPLORATION HAS GONE FROM REMOTE FANTASY TO EVERYDAY REALITY

In the stirring words of the television series *Star Trek*, space is "the final frontier"— the last unknown realm to tempt explorers, scientists, visionaries, adventurers, and perhaps even pioneering settlers. Yet the idea of traveling in space is not new; it captured people's imaginations long before the Earth was fully explored and mapped, and decades before a man stood at the South Pole. For some restless souls, the notion of space was seductive even before it truly was the final frontier.

Until relatively recently, however, most respectable scientists would have declared confidently that the idea of space travel was absurd, or that the technology for achieving it lay so far into the future that discussion was meaningless. After all, at the start of the 20th century, powered aircraft had yet to fly. And when the first one did fly, in December 1903, no one would have believed at the time that men would be standing on the Moon 66 years later.

Visionary enthusiasts

By strange coincidence, the very year that the Wright brothers made their pioneering flights near Kitty Hawk, North Carolina, a self-taught Russian mathematician published a description of a new type of rocket, specifically designed for exploring space. Konstantin Tsiolkovsky was ignored at the time, yet the propulsion system and the fuels he proposed—liquid hydrogen and oxygen—were precisely what took the Apollo spacecraft to the Moon in 1969.

Other lone experimenters forged ahead, often without knowing of each other's work. In the United States, Robert Goddard ignored insults in the media and official indifference to build successful working models of liquid-fuel rockets, while Hermann Oberth in Germany set his sights far higher, on a working space rocket, though he never achieved this dream.

At this time, in the 1920s, the world's few rocketeers were often viewed as cranks, especially if they dared to link their work with the possibility of space flight. Rocket

EARLY GENIUS The Tsiolkovsky State Museum of the History of Cosmonautics, at Kaluga, near Moscow, is named after Konstantin Tsiolkovsky, often described as the "father of modern rocketry." His thinking was years ahead of its time, and was appreciated by only a few of his contemporaries.

science was advanced by the sheer commitment of dedicated enthusiasts, such as the members of Germany's vigorous *Verein für Raumschiffahrt* (Society for Space Travel).

From these benign beginnings, however, the world's most inspired rocketeers had to enter a darker world, in which they explored their creations' capacity for destruction. Rockets have always had a dual nature—the potential to achieve wonders, or to wreak havoc. The fireworks that sprinkle colored stars over a mesmerized crowd become deadly missiles if aimed horizontally, rather than vertically. The boosters that have lifted satellites and men into orbit are cousins, if not siblings, of military rockets armed with high explosive or even nuclear warheads. It is therefore not surprising that as Germany grew more bellicose and started rearming itself, so the potentially destructive nature of rockets as weapons came to the fore.

In the early 1930s, the lonely work of German enthusiasts ended abruptly as the army took an interest in what they were doing. At last the enthusiasts had the money, facilities and encouragement to undertake major research, and by 1939 they had successfully launched a rocket weighing 1 ton. Three years later they would launch a rocket weighing 12 tons, and in 1944 this monster—named the V-2—would bring terror to London and Antwerp, Belgium, delivering its 1 ton warhead at over four times the

ROCKET MAN Robert Goddard (below, far left) and his assistants work on the fuel pumps of a rocket at Goddard's famed shop in Roswell, New Mexico, in 1940.

speed of sound. For a while, observers were mystified by the weapon's apparent double explosion, learning later that the second blast was the sound wave of the rocket catching up with it after it had arrived.

Behind these developments stood a man of genius, the young Prussian aristocrat Wernher von Braun. History might have judged him very differently, had he not decided, in the last days of the war, to offer his expertise to the Allies, specifically the Americans. As a result, he is only occasionally associated with Hitler's fearsome V-2, and is better remembered for his later work as the creative driving force behind America's postwar space program, and the Moon rocket itself.

Meanwhile the Soviet Union had failed to live up to some promising developments. In the early 1930s, the far-sighted armaments minister, Mikhail Tukhachevsky, began supporting the work of the brilliant engineer Sergei Korolev. But

SOVIET HEROES Yuri Gagarin (below left) and Sergei Korolev (below right) look suitably cheerful after Gagarin's historic first manned flight in space. The rocket that took him there in April 1961 was designed by Korolev.

when Stalin decided, in a sudden fit of paranoia, to execute Tukhachevsky, Korolev nearly shared the same fate. If he had, the subsequent U.S.–Soviet space race might not have been so hard fought. Instead, Korolev won a reprieve,

GERMAN KNOW-HOW Key members of the U.S. Army Ballistic Missile Agency in 1956 included German rocket scientists who had earlier designed the much-feared V-2. They are Hermann Oberth (front) and Wernher von Braun (sitting on the table).

was sent first to a Siberian mine and labor camp, and then made to work in an aircraft factory. Only after the war did the Soviets realize what ground they had lost, and Korolev was

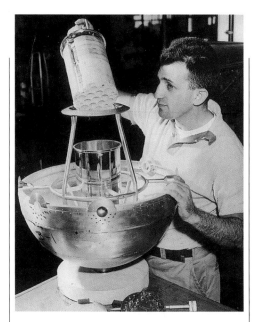

hastily reassigned to his beloved rockets. In the years that followed, he made up for lost time by building the rockets that put the first satellite, and then the first man, into orbit around the Earth.

To the Moon

For much of the decade after the Second World War, rockets continued to be developed primarily for their military potential. The V-2 had shown that rockets could outperform any artillery in terms of their range and the amount of explosive they could deliver. The development of massive explosive power in the relatively light atomic bomb spurred military planners into finding a way of mounting this device on top of a rocket, which might lob it across whole continents.

Intercontinental flight inevitably involved entering the outer reaches of the Earth's atmosphere, from where new and exciting data was being gathered. For visionaries like Wernher von Braun, intercontinental ballistic missiles (ICBMs) provided a welcome means to an end, which for him was the exploration of space itself. He therefore continued to

MAN-MADE MOON America's Project Vanguard aimed to place the world's first artificial satellite in orbit during 1957-8, the International Geophysical Year. Here, the gold-plated sphere is being fitted with its instruments. The Soviet Union's Sputnik would win this lap in the space race.

develop missiles for the Pentagon, but dreamed of space stations, a manned Moon mission, and later missions to Mars. Such thoughts may have been deferred indefinitely were it not for the intense U.S.-Soviet rivalry of the Cold War.

The Soviet Union's success in exploding a hydrogen bomb in 1953, less than a year after the United States had made the breakthrough in the technology, prompted deep unease in America. This unease intensified when the Soviet Union succeeded in being the first to place a satellite in Earth orbit in

1957, followed by the first man in 1961. The United States could not allow itself to be led by the nose, and President John F. Kennedy responded accordingly.

In a typically bold move, Kennedy committed the United States to placing a man on the Moon before the end of the 1960s, giving the country's scientists, engineers and fledgling space industry less than nine years to accomplish the task. They rose to the challenge magnificently, and on July 20, 1969, while the world held its collective breath, Neil Armstrong and Buzz Aldrin stepped onto the Moon.

Even so, until the very end, no one could be sure that the Soviets would not beat them to it again. A veil of secrecy surrounded Soviet missions, so that those that failed were never revealed to the world, while those that succeeded were trumpeted abroad as triumphs of the inherently superi-

DRESS REHEARSAL The Apollo 10 spacecraft, perched atop its three-stage Saturn V launch rocket, awaits liftoff at Cape Canaveral in May 1969. The mission flew to within 9 miles of the lunar surface in a final, tantalizing test that gave Apollo 11 the go-ahead to land on the Moon itself.

or socialist way of life. The reality was that the race for the Moon strained the Soviet space program to its limits. In developing their own technology, the Soviets had made some very significant achievements, but they simply could not compete with the sheer economic muscle of the United States; nor was it realistic for them to attempt to match Kennedy's agenda.

Having lost the race for the Moon, they therefore turned their attention elsewhere, and once again began to lead the way. While the Americans sent up another six Moon missions, one of which—Apollo 13—nearly ended in disaster after an on-board explosion, the Soviets launched the first of several space stations, beginning with Salyut 1 in 1971. The Americans followed suit with Skylab in 1973, but ended the project the following year. The Soviets, by comparison, pushed on with Salyuts 5, 6 and 7, doing

LATER APOLLOS After the near tragedy of Apollo 13, the last four Apollo missions went smoothly. Apollo 15, seen splashing down in the Pacific Ocean in August 1971 (above), was the first mission to have a battery-powered Moon vehicle, called a Lunar Rover. Astronauts traveled 21 miles in Apollo 17's Lunar Rover (right), collecting a wide variety of rocks.

research and setting new records for manned space flight. In 1986, they launched Mir ("Peace"), the biggest station yet. It would be occupied for the next 13 years, with supplies ferried up by unmanned cargo rockets, and replacement crews taking over after longer and longer periods of duty.

Valuable lessons were being learned that might one day be used on a manned mission to Mars, or an even more distant planet. For

the moment, though, only unmanned probes visited our neighbors in the Solar System, sending back astonishing data and images of Venus, Mars, Mercury and the outer planets. During the 1980s and 1990s, Earth orbit would be the limit of manned missions by both superpowers, with America taking a wholly different path from the Soviet Union.

In 1981, a new era in space travel was opened with the maiden flight of the United

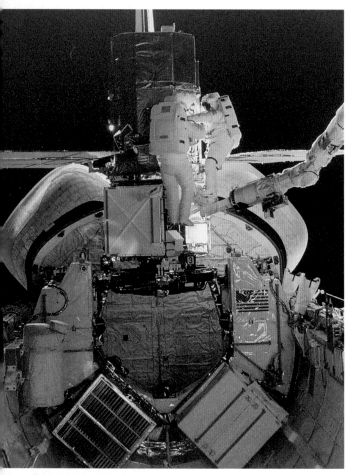

States's Space Shuttle, *Columbia*. The Shuttle, or Space Transportation System (STS for short), with its plane-like Orbiter and its re-usable rocket boosters, slashed the cost of flying into space, albeit at a relatively low altitude—the Shuttle is no vehicle for traveling to the Moon, or even deep space between the Earth and the Moon.

After only a few flights, however, the Shuttle had proved its worth, launching new satellites, capturing ailing satellites and repairing them or bringing them back to Earth, and carrying out many scientific experiments, as well as secret missions for the Pentagon. Then came the *Challenger* disaster, which killed seven Shuttle crew in January 1986. For a time thereafter only the Soviets sent humans into space.

Toward a new era of cooperation

By the time Shuttle missions resumed in late 1988, the Soviet economy and the Soviet Union as a whole were on the edge of a precipice. The U.S.S.R.'s costly space industry was one of the first to feel the effects as the country slid into economic and political turmoil. It could hardly be otherwise, since the Soviets' biggest launch site was at Tyuratam, in the Kazakh Soviet Socialist

Republic. This area of Central Asia had been conquered by the Russian tsar in the 19th century, and was quick to reassert its independence from its Soviet masters in 1991. A few years later, the great cosmodrome from which Sputnik and Gagarin were launched was a sorry scene of decay and neglect. Cosmonauts continued to work on Mir, but were plagued by endless failures and problems that could have looked like a comedy of errors if their potential consequences had not been so dire. In June 1997, the entire station looked as if it might have to be abandoned after it was punctured by an unmanned supply vessel. Two cosmonauts and a visiting American astronaut came perilously close to a horrific and untimely end.

Yet plans were already well under way for an ambitious new enterprise in space—one that would embody the antithesis of the old superpower rivalry, and point the way forward in the new millennium. At the heart of this endeavor is a giant space station, to be built with the cooperation of 16 nations. The International Space Station will weigh some 450 tons and be the size of two soccer fields. Its first module, Zarya ("Sunrise"), was launched in November 1998 from the historic cosmodrome at Tyuratam,

recently refurbished by an injection of American money. The following month, a second module, called Unity, was delivered into space by the Shuttle *Endeavor*.

Once completed, the space station will be home to crews up to seven members strong. Entrepreneurs are already talking about adding a hotel for private travelers on the Shuttle, while scientists see the station as a staging post for future missions to the planets and even the stars—all at a time when John Glenn showed that age is no bar to traveling in space. The spirit of the early rocketeers is clearly still with us, and the final frontier as tempting as ever.

TOWARD THE NEW FRONTIER

SINCE ANCIENT TIMES, HUMANS HAD LOOKED AT THE HEAVENS AS A WORLD APART AND UNTOUCHABLE. THEN SCIENCE REVEALED THAT THE HEAVENLY BODIES ARE OBEDIENT TO PHYSICAL LAWS, AND A NEW HUMAN DREAM WAS BORN. IN THE 19TH CENTURY, THAT DREAM FIRED THE IMAGINATION OF THE FIRST SCIENCE FICTION WRITERS; IN THE 20TH IT INSPIRED INVENTION AS PEOPLE REALIZED THEY COULD REACH FOR THE STARS.

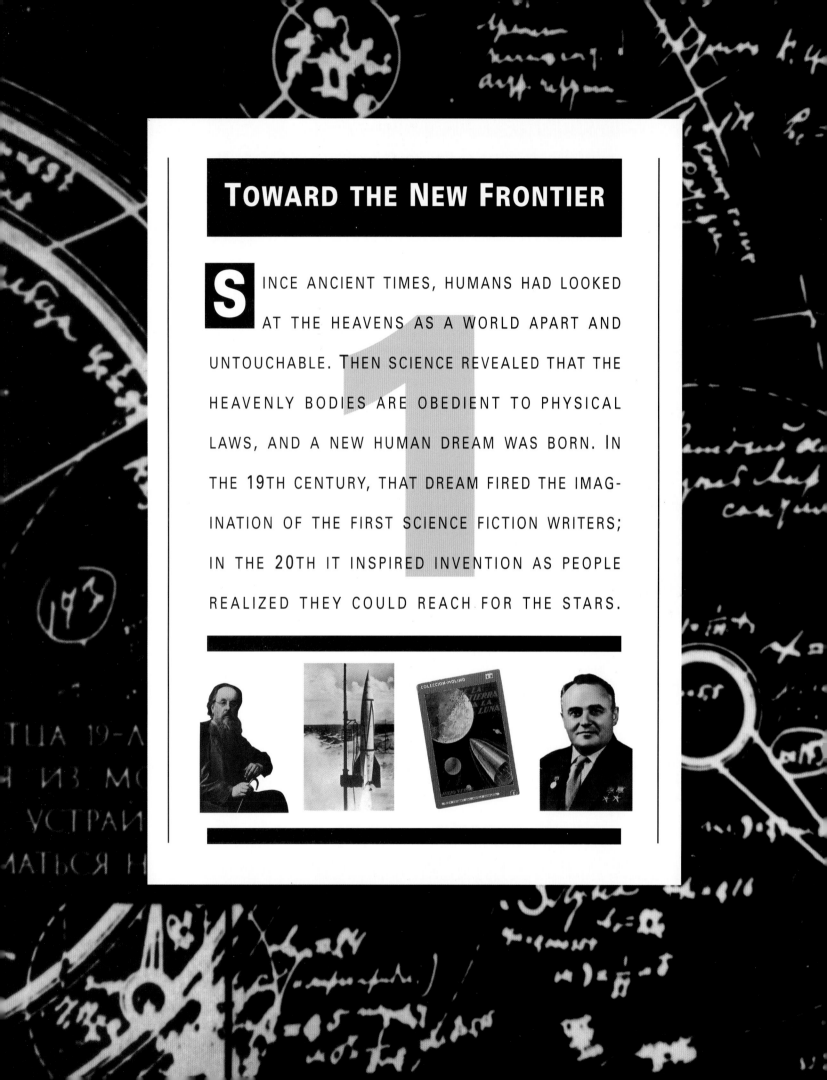

DREAMS AND REALITIES

SCIENCE FICTION AND AN ANCIENT MILITARY TECHNOLOGY—THE TWIN SOURCES OF INSPIRATION FOR THE FIRST MODERN ROCKETEERS

The conquest of space arose from two separate roots: one a dream and the other a reality. But whereas the dream of space travel was the preserve of visionaries and eccentrics until well into the 20th century, the reality that would make the conquest of space possible—the rocket—had been widely known for many centuries.

The Chinese, who invented gunpowder, were making fireworks in the 10th century. Soon afterward they attached arrowheads to their rockets, and fired them at their enemies. These "arrows of flying fire" were not especially accurate, but they succeeded in terrifying the Mongols when they invaded China in 1232. Gunpowder and rockets soon spread to India and Europe, but with the coming of guns, rockets were again reduced to mere fireworks. Only at the end of the 18th century were they resurrected as weapons of war, first in India, and then in England by the young inventor William Congreve.

Congreve's father was head of the Royal Laboratory at the Woolwich Arsenal, and

PRISONER OF MOONMEN The heroes of Jules Verne's novel *From the Earth to the Moon* found themselves contending with more than the difficulties of space travel; they also had to escape from the Moon's inhabitants, called Selenites. Published in 1865, Verne's story captured the popular imagination. Even in 1900, some scientists were arguing that the Moon might be inhabited.

THE VISION OF JULES VERNE

At age 11, Jules Verne signed up to be a cabin boy on a ship sailing for India from his hometown of Nantes, France, on the River Loire. His father hauled him back, and the young Verne promised that thereafter: "I shall travel only in my imagination." Later, as a lawyer in Paris, he did just that, sitting down before work to write stories of science-based adventure.

Later in his life, Verne would earn enduring fame as the author of *Around the World in 80 Days* and *20,000 Leagues Under the Sea*. But his first great success was a two-part novel, *From the Earth to the Moon* and *A Trip Around the Moon*. This book was serialized in 1865, inspiring lay readers and scientists alike.

Verne's imaginary spaceship now appears ludicrous: it consists of the shell of an enormous cannon. But if such a device were really fired its occupants would be crushed. It could also never overcome air-resistance or gravity. Nor could it return to Earth, as Verne describes, by plunging through the atmosphere at 115,000 mph without burning up and killing its occupants.

Even so, the scientific facts that Verne marshaled were often correct in principle. He knew that a spaceship had to reach a high speed and a certain velocity in order to escape Earth's gravity, although the speed he quotes of 54,000 feet per second is actually faster than required. Verne also gave his spaceship retrorockets—which he correctly presumed would work in space—for achieving lunar orbit and returning to Earth. The wealth of scientific data in the book gave it a spurious authenticity.

Verne's popularity was immense, infusing anyone interested in space flight with the belief that it could, and would, happen. That same faith underlay the research and writings of experimenters and theorists who, in the decades to come, would turn Verne's wild dreams into reality.

IMPACT AHEAD Having been blasted from Earth by a giant cannon, Jules Verne's imaginary spaceship approaches its destination in this cover illustration of a Spanish edition of his novel *From the Earth to the Moon*.

thus partly responsible for both British weaponry and official fireworks celebrations. In 1793, Britain and France began a long struggle that ended at Waterloo in 1815. And in a distant echo of the European struggle, Britain was also fighting the Indian state of Mysore, which had a formidable 5,000-strong corps of rocketeers. Their flaming metal tubes could travel 1,000 yards, and made a strong impression on British troops and weapons experts.

William decided to draw on his father's experience and his own childhood memories to develop a war rocket. He devised a metal tube $3\frac{1}{2}$ feet long and 4 inches in diameter, which could travel some 2,000 yards. Such rockets were cheap, easy to transport and had no recoil, which gave them a great advantage over cannons. Military commanders agreed. In 1806, salvos of Congreve rockets proved their effectiveness by setting fire to the French coastal town of Boulogne. Soon afterward, they burned down half of the city of Copenhagen.

England acquired a Rocket Brigade, Congreve had honors heaped upon him, and his rockets became an established part of

1903 Tsiolkovsky proposes using rockets for space exploration

1904 In the U.S., Robert Goddard starts experimenting

1918 Goddard demonstrates his "bazooka"

1919 Goddard asserts that a rocket could reach the Moon

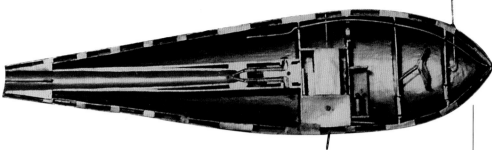

the arsenals of all Western nations. They were also immortalized in a surprising way. In 1814, during Britain's last war with the United States, Congreve rockets set the White House ablaze. Two days later, more were fired at Fort McHenry, guarding Baltimore. Watching from a British ship was an American lawyer, Francis Scott Key. His relief at seeing the American flag still flying after the bombardment inspired him to write "The Star-Spangled Banner," which refers to the "red glare" of Congreve's rockets.

For another 50 years, rockets remained a feature of many armies, until the growing accuracy, range and explosive power of artillery made them obsolete. The last appearance of war rockets was in 1881, when the Russians used them to blast their way into the Central Asian city of Geok-Tepe. At the time, no one thought of using rockets to get into space.

Flights of imagination

The dream of space travel emerged from developments in other areas, particularly astronomy and astrophysics. The invention of better telescopes in the 17th century had revealed that the Earth was part of a family of planets. This discovery suggested to a few writers that it might be possible to visit other planets in the Solar System. The means of transport, however, remained utterly fanciful. The 17th-century French writer Cyrano de Bergerac, for example, described people being transported in a ship drawn up by the heat of the Sun. Space travel remained in the province of fantasy.

Fact began to modify fantasy in the late 19th century, when another Frenchman, Jules Verne, made space travel sound plausible. He was good at using scientific facts to make wild ideas—such as traveling to the center of the Earth—seem sensible. In 1865 he published a two-part novel, *From the Earth to the Moon* and *A Trip Around the Moon*, which included a mass of scientific facts and explanations.

Verne accepted, for instance, that the biggest problem for any would-be space traveler is to travel fast enough to escape Earth's gravity. He dealt with this by having his characters devise the world's biggest gun, to fire a spaceship in the form of an artillery shell. Of course, no known material, let alone a human being, could withstand such an explosive rate of acceleration, but the idea was presented so convincingly that Verne's

PRESCIENT DESIGN Tsiolkovsky's spaceship was never built or launched, yet it was based on sound principles. It included chambers for liquid oxygen and hydrogen, the fuels that would eventually take rockets into space.

novel became not only popular but respectable. Even scientists talked about Verne's ideas, and many young minds found inspiration in his pseudoscience.

Verne's approach was later adopted by H.G. Wells, another writer fascinated by the power of technology. His novel *The First Men in the Moon* (1901) was popular for similar reasons, was similarly inspirational, and had an even less practical foundation. Wells's spaceship was driven by antigravity material called Cavorite, which masked an insolvable problem.

Tsiolkovsky's work on rocket fuels

In the history of space travel, dreams first began to acquire a sound theoretical base in the mind of a shy Russian mathematics teacher, Konstantin Tsiolkovsky. The son of a forester, he was left deaf by a childhood attack of scarlet fever. His deafness turned him into a retiring and bookish teenager who, through his studies, rose above his lowly background.

Tsiolkovsky's early work on the properties of gases and light brought him to the attention of leading scientists, and won him a small grant from the Academy of Sciences. Later, having attained a good grounding in astronomy, mathematics and engineering, and

VILLAGE TEACHER Tsiolkovsky had to wait until old age before winning the respect of the Soviet authorities. He and his work—such as these complex calculations—are now honored by a museum in the village of Kaluga, where he taught for most of his life.

1926 Goddard launches the world's first liquid-fuel rocket

1927 Formation of the Society for Space Travel; Oberth is invited to join

1928 Willy Ley publishes *The Possibility of Space Travel*

1929 Fritz Lang's film *Woman in the Moon* uses a model rocket designed by Oberth

1933 Tsander and Korolev test their rockets in Russia

1937 During Stalin's purges, Korolev is arrested and forced to work on aircraft design

secure in a teaching post in Kaluga, some 85 miles south of Moscow, he went on to become the first person to think through the problems involved in conquering space.

The conclusions that Tsiolkovsky arrived at during the 1890s were to prove remarkably accurate. He opted for rocket power over other forms of propulsion, but knew that current fuels based on black powder—a variation of gunpowder—had nowhere near the exhaust velocities to lift a rocket into space. He devised formulas to correlate speed, rocket weight, fuel weight and fuel flow. His chemical and mathematical expertise led him to suggest new fuels, including oxygen and hydrogen. He saw that these gases, when liquefied and compressed by extremely low temperatures, would then expand on ignition with enough force to achieve space flight. In these matters, Tsiolkovsky showed astonishing prescience. Liquid oxygen, which expands some 900-fold when it turns to gas, would become a standard fuel in rocketry. Liquid hydrogen, first produced in 1898, was eventually used as a rocket fuel in the 1960s.

Tsiolkovsky's first article, "Exploring Cosmic Space by Means of Reaction Devices," was accepted by a Russian magazine, and the first installment was published in 1903. The magazine was then closed down, and the article's second installment did not appear until 1911. Understandably, the piece was totally ignored. War and revolution curbed Tsiolkovsky's research. Only in the 1930s, during the last ten years of his life, was he acknowledged by the Soviet authorities as the father of rocket research. His ideas then became more widely known, inspiring a Latvian engineer, Fridrikh Tsander, to found a club for the study of astronautics. One of its members, Sergei Korolev, would one day head the Soviet space program.

On May 1, 1935, at the age of 77, Tsiolkovsky addressed May Day marchers in Moscow's Red Square by radio. He had lost none of his enthusiasm, and conjured up a vision for his audience of rockets being sent into space from Soviet cosmodromes. He died in September the same year, believing that the Soviet government was about to turn his dreams into reality.

By this stage, many of Tsiolkovsky's ideas had been arrived at independently by two other pioneering space scientists: Robert Goddard in America, and Hermann Oberth in Germany.

Goddard's liquid-fuel breakthrough

Goddard was a modest New Englander with a pioneer's ingenuity and a dogged ambition to research the upper atmosphere. Through his years as student, teacher and physics professor at Clark University in Worcester, Massachusetts, he tinkered with rockets, planning to use them to carry instruments into the stratosphere.

In his first experiments, conducted between 1904 and 1916, Goddard strapped his rockets to test benches to try out his ideas about fuels, combustion chambers and nozzles. He also tested the idea that a rocket would create thrust in a vacuum. He knew, from the first principles of physics, that it should, but there were some who held that a rocket needed to "push against" air to move forward. Goddard needed experimental proof that they were wrong. He obtained this proof by building a pistol into a small chamber, pumping out all the air, and then firing a blank cartridge. The gun recoiled in the vacuum as it did under

WORKING MODEL American rocket pioneer Robert Goddard shows off one of his first little rockets, built in 1918. At this point Goddard was still experimenting with different types of solid fuel.

normal air pressure. Further tests showed that rockets actually operate more efficiently in a vacuum, since they did not have to overcome air resistance.

Such work cost money and Goddard needed help, to the tune (he guessed) of $10,000. To obtain the money, he wrote up his findings in a paper, in which he laid down the rules governing exhaust velocities, weights of fuels and rocket-casings, heights to be achieved, the problem of air resistance, and payloads. For reaching high altitudes, he suggested the use of a "step-rocket," made up of two or more stages mounted one on top of another. Each stage would contain its own rocket-propulsion system. The lowest stage would fire first, and when its fuel was exhausted it would drop off and the next stage would fire. As each stage would benefit from the speed attained by the previous stage, each would need less fuel and therefore carry less weight than preceding ones.

After several rejections, his paper was finally accepted by the Smithsonian Institution in Washington. When asked how much cash he would need to continue his work, Goddard didn't dare tell the truth, in case the Smithsonian turned him down com-

pletely; he asked for $5,000. What he received was a check for $1,000. He had few options but to accept and go to work.

The next stage of Goddard's research led him to look at ways of controlling a rocket's rate of acceleration in order to allow a slow takeoff through the dense air of the lower atmosphere, with faster and more effective acceleration as altitude increased and air resistance decreased. Then in 1917, the United States entered the First World War, and the government gave Goddard a small grant to develop rockets for war. In 1918, he demonstrated one of his new devices, a projectile 4 feet and 3 inches long and about 2 inches wide, which could be fired from a tube. It was an early type of bazooka, and it worked perfectly. But the war ended, the weapon went into storage, and Goddard returned to his own research.

In 1919, he published a report on his work, blandly titled *A Method of Reaching Extreme Altitudes*. It contained the startling conclusion that it was possible to build a rocket that could attain the velocity of 25,000 mph needed to escape the Earth's gravity and "fall onto the moon." To Goddard's astonishment, this throwaway line

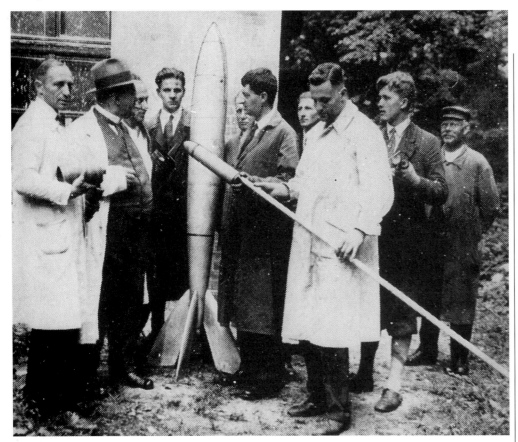

won him instant fame. The *New York Times* used the story on its front page under the headline: BELIEVES ROCKET CAN REACH MOON.

The next day, however, an editorial in the the same newspaper ridiculed Goddard, accusing him of deliberate deception. The writer claimed that Goddard must know that a rocket needed "something better than a vacuum against which to react," and concluded with the libelous innuendo that "There are such things as intentional mistakes." From this and other attacks, Goddard gained a reputation for being "moon mad." He responded with restraint, and quietly went on with his work.

Dissatisfied with solid fuels, Goddard came independently to the same conclusions as Tsiolkovsky, although he did not know the latter's work. He began to experiment with liquid fuels, first gasoline and liquid oxygen, each with its own tank. In his first liquid-fueled rocket motor, he incorporated a third tank, containing compressed carbon dioxide, to force the other two liquids into the combustion chamber. Finding that this method did not work well enough, he began designing a motor in which the liquid oxygen itself provided the necessary pressure.

On March 16, 1926, Goddard's rocket was ready for testing. It was unusual in consisting of a slender frame 10 feet long, with its motor in the front and two fuel tanks behind. Goddard and his new wife, Esther, along with two university staff members, carried the spidery rocket and its stand out into a frozen field on a farm owned by his aunt, Effie Ward. He set it up, opened the fuel valves and lit the vapor with a blowtorch fixed to the end of a pole. With a roar, the rocket lifted off. Two-and-a-half seconds later—Goddard was timing the flight with a stopwatch—the rocket exhausted its fuel and tumbled to earth 184 feet away. No public announcements were made of this, the world's first liquid-fuel rocket flight.

Experiments at Roswell

Over the next three years, Goddard refined his invention, placing the motor at the rear and devising more efficient fuel feeds. In July 1929 he tried again with his improved design, which now contained a barometer, a thermometer, a camera and a parachute. It rose to 90 feet, turned, flew 171 feet more, and landed safely, but not secretly. Headlines in the *New York Times* trumpeted: METEOR LIKE ROCKET STARTLES WORCESTER, to the horror of the local fire marshal, who promptly banned all further flights in the area.

Future research was guaranteed, first by the army, which offered Goddard the use of a local artillery range, and then by Charles Lindbergh, who had recently made the first solo flight across the Atlantic. Having read of Goddard's work, Lindbergh used his immense popularity and his expertise as an aviator to lobby for money from the millionaire philanthropist Daniel Guggenheim. A generous grant provided Goddard with a two-year sabbatical to pursue his work.

Goddard chose to go to Roswell, New Mexico, an area that was free of restrictions and had a congenial climate all year round. Here he refined the rocket's motor and began to devise a proper steering system. To do this, he added vanes that acted on the exhaust like rudders and were controlled, through pistons, by gyroscopes. In 1932, his next design flew briefly, and crashed. By then, the two years were up and he returned to Worcester, until the Guggenheim grant was renewed in 1934.

Over the next few years, Goddard tackled other crucial problems presented by his increasingly powerful rockets. The most demanding was the paradoxical requirement to create as much heat as possible while keeping the combustion chamber from melting. This demanded the use of heat-resistant materials and/or coolants, and more original thinking. No material was heat-resistant enough, and any conventional cooling method, such as a water jacket or fins for radiating heat, would add to the rocket's weight. More fuel would then be required to accelerate it to the required velocity, thus compounding the problem. He decided that the only solution was to use the rocket's own fuel as a coolant by directing it around the outside of the combustion chamber.

This and dozens of other ideas fell on deaf ears in the United States, where there was no official interest in rockets until 1940, by which time Germany was well ahead in rocket research. Even then, the authorities took note of Goddard only when a postgraduate researcher, who had seen the 1918 "bazooka" demonstration, sent a report of his work to defense officials in Washington.

Soon after, the National Defense Research Committee set up a rocket-research section, which employed Goddard during the war. He died in August 1945, four days before the Japanese surrendered.

Oberth dreams of space travel

Germany owed its start in rocketry to one man, Hermann Oberth, who worked on the same problems as Goddard and came to the same conclusions. Neither man had any notion of the other's progress until 1922, when Oberth came across a reference to Goddard's 1919 article and sent away for a copy. Scanning it eagerly, he realized that his own theories were far more ambitious than Goddard's. If anyone could claim to be the true father of modern rocketry, it would be Oberth. Tsiolkovsky, Goddard and Oberth all sowed seeds, but only Oberth's blossomed. His work led directly to the V-2 rocket, whose chief engineer, Wernher von Braun, would later create the Saturn V rocket that would take the first men to the Moon.

Oberth was born in 1894 in Transylvania, which was then a province of Hungary. Of German descent, he went to Munich to study medicine. Fascinated by the works of Verne and the German science-fiction writer Kurd Lasswitz, he signed up for two subsidiary courses, in mathematics and astronomy. In 1914, he was called up as a physician in the Austro-Hungarian army. His sector saw little action, and he used the time to work out the mathematics of space travel. In 1918, Oberth returned to Munich and abandoned medicine for astrophysics. He was almost ready to publish his ideas in 1922 when he heard about Goddard and anxiously read his article. Then, relieved that he had not been pre-empted, Oberth had his own paper published.

His ambitions soared past Goddard's—no mere missile to the Moon for him. In his 1923 book, *The Rocket into Interplanetary Space*, Oberth asserted that space rockets will not merely "be able to leave the zone of terrestrial attraction," but that "such machines ... will be able to carry men," and do so with profit, "within a

THE WOMAN IN THE MOON

When Fritz Lang's silent film *Frau im Mond* (*Woman in the Moon*) appeared in 1929, Lang already had an international reputation. His earlier works included *Doktor Mabuse*, a melodrama about a power-crazed hypnotist, and *Metropolis*, a disturbing futuristic vision; both had won instant recognition. Lang was therefore well placed to explore a new obsession: space travel.

In 1927, Lang met the science writer Willy Ley, a founding member that year of the *Verein für Raumschiffahrt* (Society for Space Travel). When Lang started to plan a film about flight to the moon, Ley suggested that he should use the rocket scientist Hermann Oberth as his technical adviser. Oberth joined the team, working closely with Ley and Lang on designing the sets, and advising Lang's wife, Thea von Harbou, who was writing the screenplay.

The story concerns Wolf Helius, a scientist, who is also the rocket's pilot; his chief engineer, Hans; Hans's fiancée, the brainy female astronomer Frieda; and a stop-at-nothing financier, Walt Turner, who is keen to get his hands on the Moon's gold. Frieda was played by Gerda Maurus, the star who was Lang's mistress in a very public affair.

Besides Gerda Maurus, the film's major attraction was its special effects. Lang was a stickler for facts, which Ley and Oberth provided. As a result, the rocket's launch, almost halfway through the film, is the high point. There then follows a plot that now seems ludicrous: the two spacemen in love with Frieda, gold discovered in lunar caverns, hero and heroine marooned by the departing spaceship to face a slow death.

German critics gave the film muted praise. When it arrived in the United States in 1931, American critics panned it as a bore. For one thing, it was a silent movie and sound had just made its debut. Yet the film has retained a mystique because of its technical wizardry and prescience. Lang's Moon may have a breathable atmosphere, but other details are wholly acceptable today, such as the multistage rocket and the backward countdown, which became a staple of blastoff protocol everywhere.

After the Nazis came to power in 1933, Lang fled to Hollywood, where he became a legendary figure. He always retained an affection for *Frau im Mond*, and was delighted to be told, at a space-science seminar in 1968, that he was considered to be one of the "fathers of rocket science." He died in 1976, having seen his dream of lunar conquest come true.

CONVINCING DESIGN Hermann Oberth's concept of a two-stage rocket for *Frau im Mond* was not only scientifically accurate, it was in many respects ahead of its time, anticipating developments that were yet to come.

STYLISH BLAST-OFF To advertise his film, Lang used Oberth's rocket to create an Art Deco image. Lang's wife wrote the script, and his mistress, Gerda Maurus, was the star.

BRIEF FLIGHT In November 1932, a group of German enthusiasts met at Berlin's Tempelhof airfield to launch an experimental rocket. Fitted with huge stabilizing fins (right), the rocket blasted off successfully (above).

few decades." The treatise that followed consisted mainly of a mathematical thesis, but it ended with descriptions of a space suit and how an astronaut might move about in weightless conditions. Oberth even suggested that a 400-ton rocket could carry two astronauts into orbit, where they could build an orbiting space station that "would see every iceberg and warn every ship."

Most scientists ignored the book, as did businessmen who might have invested in Oberth's research. But it struck a popular chord, and it inspired a dozen men in Breslau (today the Polish city of Wrocław) to found the world's first rocket society, the *Verein für Raumschiffahrt* (Society for Space Travel), or VfR, in 1927. Oberth was invited to join, as were several other eminent scientists from Russia and France, as well as a young enthusiast named Wernher von

Braun. The following year, a young science writer named Willy Ley edited a book with the ambitious title *The Possibility of Space Travel*. It included contributions by several scientists, including Oberth.

Yet the real impetus to Oberth's work came from a surprising quarter: Germany's most famous film director, Fritz Lang. In 1926 Lang had used robotics as the theme of his classic silent film, *Metropolis*; a year later

he was thinking about space travel. Lang's proposed new film, *Frau im Mond* (*Woman in the Moon*), was going to be a romantic fantasy about a trip to the Moon, and Lang wanted the spaceship to look authentic. To achieve this, he invited Oberth to be his technical adviser.

Oberth responded by designing a very convincing mock-up, complete with a cabin and a parachute to ensure a safe return. He then asked Lang for funds to build a real rocket. Lang, with an eye on publicity, talked his film company into agreeing. Oberth was free to begin work, which he did with two assistants, whom he recruited by placing advertisements in newspapers.

Aiming to have the rocket ready for the film's launch, Oberth began a series of desperate experiments, all of which showed him the impossibility of his task. Running into debt with Berlin metalworkers, he fled in shame to Transylvania. Returning for the film's premiere, he was a diffident and shamefaced figure, removed from the glitterati who arrived in limousines to see his visionary Moon rocket on the screen.

Oberth never reaped any benefits from his originality, nor from his growing reputa-tion. He was awarded a French prize of 10,000 francs for achievements in astronautics, and he published a second book, which was well received. Yet neither supplied enough cash to build rockets. With a family to support and a teaching post open to him in Transylvania (now part of independent Romania), Oberth had no option but to return home. There he remained until 1938,

FIRST PRINCIPLES OF ROCKETRY

All engines create energy by combustion, and then convert that energy into motion. Combustion requires oxygen, and most engines draw their oxygen from the air. Forward motion is provided in one of two ways: either by direct contact with surrounding matter, as in the case of a car's wheel or a ship's propeller; or by reaction, as in the case of a jet engine or a rocket.

In a reaction motor, forward motion is produced when exhaust gases are ejected. This is described by Newton's third law of motion, which states that every action has an equal and opposite reaction. It is the force that causes a garden sprinkler to spin as it ejects water, and a gun to recoil when it is fired. It drives a jet forward and carries a rocket aloft, whether it is simple fireworks or a space shuttle.

Whereas a jet engine draws its oxygen from the air, a fireworks rocket and a space rocket both carry their own oxygen supply. In the case of fireworks, the oxygen is locked into the saltpeter, a constituent of gunpowder. In a space rocket, liquid oxygen is a component of the fuel.

In a rocket, burning fuel causes a slow-motion explosion inside a combustion channel or chamber, converting the fuel into gas, which then escapes through the exhaust nozzle. As exhaust gases escape, they drive the rocket in the opposite direction. If the thrust of the exhaust is powerful enough, it will overcome the rocket's inertia and lift it off the ground.

If a rocket engine is fired in space, its thrust can be used in a number of ways, depending on where the exhaust is pointing. The thrust can propel the spacecraft forward and accelerate its speed; it can be used to steer the spaceship; and retrorockets can be used to slow or even stop the spacecraft.

At the turn of the century, few people understood these principles. It was widely believed that a rocket needed something, such as air, to push against, and so could not work in space. It was not until the 1920s that rockets were accepted as the only way to achieve space flight.

SOLID AND LIQUID **The three most common types of rocket are the fireworks rocket (right), the solid-fuel rocket (center) and the liquid-fuel rocket (far right). They all share the same basic principles. Fuel burns in a combustion chamber or channel, creating pressurized gas. The gas escapes through an exhaust, exerting a force that creates an equal and opposite reaction, thereby driving the rocket forward (below).**

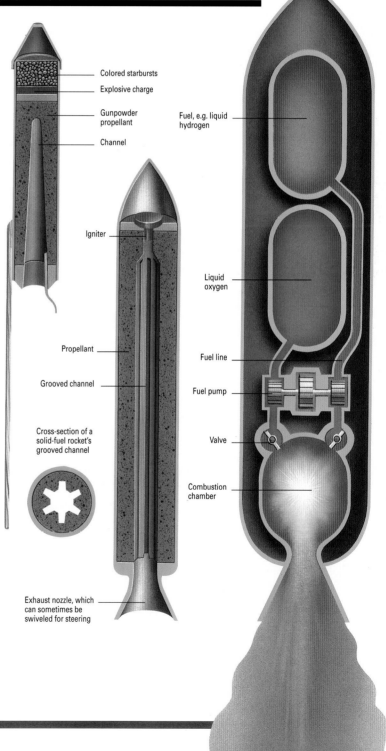

Colored starbursts
Explosive charge
Gunpowder propellant
Channel
Fuel, e.g. liquid hydrogen
Igniter
Liquid oxygen
Propellant
Fuel line
Grooved channel
Fuel pump
Valve
Cross-section of a solid-fuel rocket's grooved channel
Combustion chamber
Exhaust nozzle, which can sometimes be swiveled for steering

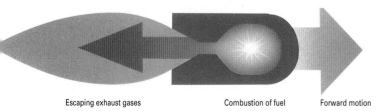

Escaping exhaust gases   Combustion of fuel   Forward motion

when the College of Engineering in Vienna offered him a post on a rocket research program. He didn't know it, but the offer concealed a hidden agenda. By then, Austria had become part of Hitler's Third Reich, and war was brewing. As a Romanian, Oberth might have been open to British offers, so the Reich moved first and Oberth accepted.

Transferred to an army project in Dresden, he soon found himself so confined in his research that he declared that he was going home. Then he discovered the truth: he knew too much to be allowed home. Fearing imprisonment, Oberth applied for German citizenship. When this was granted, he was promoted to the Reich's recently established rocket research station at Peenemünde on the Baltic coast.

Yet again, Oberth found himself on the sidelines, having to work under the brilliant young engineer and astrophysicist Wernher von Braun. Oberth never emerged from obscurity, settling after the war in the United States, before finally returning to Germany.

Developments in the Soviet Union

In the meantime, rocket science had undergone a double revolution in the Soviet Union. At first, in the early 1930s, the authorities had built on Tsiolkovsky's heritage; then, just two years after his death in 1935, they betrayed it.

By the early 1930s, a successful rocket research group was being run in Moscow by two scientists, Fridrikh Tsander and Sergei Korolev. Its work came to the attention of the army's young armaments minister of the time, Marshal Mikhail Tukhachevsky, who was fascinated by rockets as weapons, as motors for airplanes and as possible space vehicles.

With a wary eye on the work of Goddard in the United States and Oberth in Germany, Tukhachevsky took over control of the Moscow group, funding it through the army. In 1933, the group's first test rockets took briefly and erratically to the air. Korolev became an officer in the Red Army; he also wrote a book, *Rocket Flight in the Stratosphere*, in order to promote his ideas. Then, just as Soviet rocketry was edging forward, catastrophe struck.

In 1931, Joseph Stalin had declared that the Soviet Union was to make up for a century of stagnation in a decade of concentrated effort. Industry was to be transformed and all agricultural lands brought under state control. Some 25 million people were forcibly transported from the countryside to towns. Amid immense suffering, opposition was crushed in what came to be called the Great Terror.

In this climate of paranoia, Stalin then turned on his perceived "enemies" in the Communist Party and the Red Army. In June 1937, Tukhachevsky and eight of his

LOST OPPORTUNITIES Mikhail Tukhachevsky, the Red Army's far-sighted armaments minister, was an enthusiastic supporter of Soviet rocket research until his death in 1937 during one of Stalin's many purges.

SECRET GENIUS Sergei Korolev inspired and guided Soviet rocketry for 30 years, yet few people ever knew about him. His name was kept secret, and official documents referred to him simply as "the chief designer."

colleagues were accused of treason, and summarily shot.

All of Tukhachevsky's projects, and their commanders, were instantly at risk. Two rocket researchers were executed. Korolev was arrested and sentenced to ten years' slave labor in the Kolyma gold mines of Siberia. In the typical manner of the Soviet secret police, the NKVD, there was no investigation, merely an accusation. "Our country does not need your fireworks," his interrogator said. "Or maybe you are making rockets for an attempt on the life of our leader?"

As it happened, the NKVD was also purged, and Korolev's sentence was altered. He was brought back to Moscow and assigned to a group of aircraft designers who were being made to work in prison under the top Soviet aeronautical designer, Andrei Tupolev. And that, for the time being, was the end of Soviet rocketry, which might, in a different world, have challenged the commanding lead now enjoyed by the Germans.

THE INSPIRATION OF WAR

GERMANY'S PRE-EMINENCE IN ROCKETRY CONTRIBUTED LITTLE TO ITS WAR EFFORT, BUT LAID THE BASIS FOR THE AMERICAN SPACE PROGRAM

THE FLIGHT OF REPULSOR NO. 1

The science writer and rocket pioneer Willy Ley describes how, on May 14, 1931, he and a colleague ignited Repulsor No. 1, which looked like an egg bolted between two stilt-like fuel tanks.

The thing took off "with a wild roar … It hit the roof of the building and raced up slantwise at an angle of 70 degrees. After two seconds it began to loop the loop, spilling all the water out of the cooling jacket, and came down in a power dive. While it was diving the wall of the combustion chamber—being no longer cooled—gave way in one place, and with two jets twirling it, the thing went completely crazy. It did not crash because the fuel happened to run out just as it pulled out of a power dive near the ground. Actually, it almost made a landing. Examination showed that it was intact save for the hole in the motor … we were dizzy from watching and jumping out of the way, and had to sit on the grass for a moment."

In the early 1930s, Germany seized the initiative in rocket research as a direct result of Hermann Oberth's work sponsored by the film director Fritz Lang. When Oberth advertised for help in building a working rocket, one interviewee was an ex-combat pilot, engineer and burglar-alarm salesman named Rudolf Nebel. Oberth hired him instantly. When Lang's money stopped, Nebel continued rocket research on his own as a member of the expanding VfR (Society for Space Travel).

Knowing that bankers and businessmen could not be expected to provide funding for his work, Nebel approached the German army. By chance, the chief of research and development for army weapons, General Karl Becker, saw a military use for rockets. According to the Treaty of Versailles, which had imposed draconian peace terms on Germany in 1919, the country could not develop heavy artillery. But the treaty made no mention of rockets. Becker had already written about their potential as a substitute for artillery, so he provided the VfR with a grant of $1,200 and an abandoned army site outside Berlin as a base.

Here, in their *Raketenflugplatz* ("Rocket airport"), dedicated enthusiasts such as the aristocratic 19-year-old Wernher von Braun and the science writer Willy Ley began to tinker with equipment and exchange ideas. They had little cash for research, especially after Becker withdrew funding because of Nebel's fondness for self-promotion. So they begged for materials from manufacturers, and experimented with tiny water-cooled rocket engines. In May 1931, their Repulsor No. 1 (they took its name from the rocket built by Martians in one of Kurd Lasswitz's science-fiction novels) made a brief, wild flight. Other flights followed, with increasing success, until January 1934, when the Nazis threw them out and turned their rocket port back into an ammunition dump.

Meanwhile, General Becker had started his own research program at the army proving grounds at Kummersdorf. He was determined to build a rocket that would outperform the largest guns, and delegated the task to a young officer newly qualified as an engineer, Captain Walter Dornberger. With army funds at his disposal, Becker then made an irresistible offer to one of the brightest members of the VfR, Wernher von Braun.

The aristocratic scientist

Von Braun was from a proud Prussian family. His father, Baron Magnus, was a banker and minister of agriculture in Germany's postwar Weimar Republic. From his mother, young Wernher inherited an interest in astronomy, which led him to study engineering, an unusual choice for a young aristocrat at the time.

In 1932, this tall, fair-haired student was still only 20, but already his intelligence, energy and commitment had given him an astonishing grasp of astronautics. He joined Becker's rocket program as its top civilian specialist while still a postgraduate student at Berlin University, working on liquid rocketry. Doctoral degrees often took up to seven years to complete; von Braun, working through the dramatic period of Hitler's seizure of power, received his after a mere 18 months, for a thesis whose very title was top secret. His studies provided him and his army superiors—who were now confident of a bright future through Nazi rearmament—

THINGS TO COME   Rudolf Nebel, in a typically self-confident pose, holds aloft a model of Hermann Oberth's rocket, an ancestor of the wartime V-2. Nebel fell from favor because of his irritating propensity for self-promotion.

TESTING, TESTING An A-3 rocket, 2 feet high, stands on its test pad at Kummersdorf, south of Berlin, in 1936. Most of the A-3 tests were not particularly successful.

with the theoretical basis for the next major advances in rocketry. Yet many practical difficulties remained to be overcome, as von Braun soon discovered.

From theory to practice

Progress did not come easily. In 1933, a new rocket known as Aggregat-1 (or A-1), fueled with alcohol, blew up on ignition. Clearly, new research into fuel was needed, together with gyroscopic guidance systems, air resistance and cooling methods. In late 1934, the research paid off when two prototype A-2s lifted off and soared over $1^1/4$ miles above the North Sea island of Borkum.

By 1936, von Braun had laid plans for an A-3 and an A-4, a rocket that he believed would carry a ton of high explosive some 160 miles—twice the range of any artillery piece. Such a device needed its own facilities, with space for long-range testing in secret. Von Braun himself found the place, near Peenemünde, at the northern tip of Usedom on the Baltic coast, where his father used to go duck shooting. Here, with a budget of $2.6

million, von Braun set about realizing his dreams as technical director of the new Army Experimental Station.

Growth of the base was fast, driven by the army's demand for a mobile weapon that would outperform any artillery gun and the rockets of any other nation. All advanced nations—France, Italy, Britain, Russia, the United States—were funding rocket research, mostly into short-range battlefield weapons. The most famous of these was the 7-foot Russian Katyusha, fired from a rack that became known as "Stalin's Organ Pipes." Von Braun, turning a blind eye to the destructive implications of his work, put aside his ambitions to build a space rocket for the time being.

As Hitler took over Austria, followed by the Sudetenland and then the remainder of Czechoslovakia, the army's demands for a practical weapon intensified. In 1939, a new

rocket weighing 1 ton rose 5 miles in 45 seconds. Its momentum then carried it on upward, until a parachute opened to float it safely back to Earth.

But the A-4 would be a very different machine. It was to be 46 feet in length, weigh 12 tons and be capable of accurate parabolic flight over 200 miles. It would have a motor with a thrust of $24^1/2$ tons, and a cooling system to keep the nozzle from melting. This was easier said than done. Indeed, there were so many problems to be solved that an additional rocket, the smaller A-5, was built alongside the A-4 to test many

CLOSE TO SUCCESS Dr. Wernher von Braun poses with a model of the A-4, which would later be designated V-2, from the code name *Vergeltungswaffen*, meaning "weapons of retaliation." An A-4 blasts off (below) from its launch pad test at Peenemünde in 1942.

1931 German enthusiasts launch Repulsor No. 1

1932 Wernher von Braun joins the official German rocket program

1934 Successful flight of two prototype A-2 rockets

1936 Von Braun sets up a new rocket research station at Peenemünde

experimental systems, which could then be incorporated into the A-4 if successful.

Von Braun needed a way of delivering fuel from the rocket's fuel tanks into the combustion chamber at a rate of 275 pounds per second. Clearly, he needed a pump. But what would power the pump without adding too much weight? The answer lay in chemistry, not mechanics. Von Braun used hydrogen peroxide, an unstable liquid that decomposes rapidly into hot gas when combined with manganese salts. If allowed to escape through a nozzle, the gas became a high-pressure jet of steam, capable of powering a fuel pump, and thus doing away with the need for a turbine.

Guidance and control presented whole new sets of problems. When a rocket is being propelled by its motor, it is balanced on its exhaust, with about as much stability as a broomstick balanced upright on your hand. Every tilt, whether due to gusts of wind or the rocket's own instability, has to be counteracted instantly. Von Braun set up gyroscopes that could sense such alterations, and then correct them by deflecting rudder vanes in the exhaust (a system that robbed the rocket of up to 17 percent of its thrust). He also had to devise a way of tilting the rocket gradually away from the vertical in order to achieve long-range flight.

When its fuel ran out, the rocket needed to be traveling at exactly the right speed and angle to continue its parabolic flight to its target. These issues were extremely difficult to calculate because while the rocket was using up its fuel, its weight was constantly decreasing and its center of gravity shifting. Such issues stimulated von Braun's staff to develop some of the first computer systems.

Hitler demands "an annihilating effect"

Success came slowly, for Nazi successes in the early stages of the Second World War suggested that tanks, artillery and planes were enough to achieve Hitler's ambitions, and funding was cut back. Nevertheless, test flights started in June 1942. The first A-4 crashed about a mile from its launch pad. In August came the first rocket flight to break the sound barrier. In October, an A-4 flew some 125 miles, and reached an altitude of 70 miles. But over the following six months, only one out of 11 launches did as well. Nevertheless, by the spring of the following year, the A-4 was ready. In July 1943,

INTO INNER SPACE

The 46 foot long, 12 ton A-4 rocket ready for firing at Peenemünde on October 3, 1942, had a special significance. If it failed, it would be the last that the rocketeers headed by Walter Dornberger and Wernher von Braun would have the money to build. Not only would Germany's investment in a potent weapon have been wasted—it would end their dreams of space travel.

The 10 minutes before firing were agony for those watching, and the final 60 seconds—"X minus one"—were so tense that the staff referred to it later as the "Peenemünde minute."

Finally the loudspeaker announced "Ignition!" and the A-4's nozzle blasted out spark-streaked clouds, which turned to a flame of red and yellow. For three seconds the blast remained stable, taking the weight of the rocket. Then the power increased and the rocket slowly rose, balancing on its exhaust. As the thrust built to 25 tons, the rocket accelerated upward.

After 4½ seconds the rocket began to tilt; after 20 seconds, it was traveling at 650 mph.

"Sonic speed!" the loudspeaker announced, over the rocket's fading roar.

A trail of white flared from the rear.

"The fins have come off!"

"She's falling!"

But it was merely the vapor trail. The rocket vanished to a dot and the trail developed into a zigzag created by air currents as the missile reached 1,500 mph, then 2,000 mph.

With everyone in the crowd watching through binoculars, rigid with tension, the 10 tons of fuel burned steadily for 50 seconds. Then, with the invisible rocket traveling at 3,500 mph, came the moment known as *Brennschluss* ("burning end"), when the fuel ran out.

Invisible, and completely silent except for its radio signal, the rocket soared on, reaching a height of 70 miles, halfway to space, where air pressure is reduced 100,000-fold. Then it started dropping, accelerating again, until rising air pressure raised its surface temperature to 1,250° Fahrenheit. All of a sudden, the radio signal went dead. The rocket had crashed down-range, some 125 miles away.

That evening, Dornberger announced to his jubilant staff: "We have proved rocket propulsion practicable for space travel." It was the opening of a "new era of transportation."

Dornberger was also aware that it was the start of a new era of weaponry. The A-4, soon to be renamed the V-2, would become one of the most feared weapons of the Second World War.

BREAKTHROUGH   The first truly successful A-4 lifts off from Peenemünde on October 3, 1942, on a flight that would take it to an altitude of 70 miles and the previously unheard-of speed of 3,500 mph.

Dornberger and von Braun presented Hitler with the news.

By now, Nazi ambitions had given way to desperation. On the Eastern Front, German forces were reeling after defeat at Stalingrad. British and American bombers were flattening the heavy industries of the Ruhr region. Hitler began to place his hopes in new weapons: a jet fighter, a rocket plane, a jet-propelled "flying bomb" and von Braun's A-4 rocket. When a special commission checked out the relative benefits of the flying bomb and the A-4, it concluded that both had to go into production, for both promised to carry the war back to London. The flying bomb was cheap, but slow and vulnerable; the A-4 was expensive, but it could be launched anywhere and was undetectable in flight.

When Dornberger and von Braun were summoned to brief Hitler, they told the weary Führer that the A-4 could carry a 1 ton warhead. Hitler asked if this could be

| 1939 A-3 weighing 1 ton flies 5 miles | 1942 First flight of the A-4 rocket weighing 12 tons | 1943 A-4 is ready for military use; it is renamed V-2 and production begins | 1944 First V-2s launched against London | 1945 Von Braun surrenders to the Americans; the U.S. Army captures 100 V-2s |

raised to 10 tons. Dornberger said it was possible, but would take years to develop.

"But what I want," said Hitler, eyes wild, "is annihilation—an annihilating effect!"

When Dornberger said hesitatingly that they had not thought about anything like that, Hitler shrieked: "You? No, you didn't think of it! But I did!"

Since Hitler seized upon these two weapons as miracle-workers, they were given the code name *Vergeltungswaffen* ("weapons of retaliation"). The flying bomb would be the V-1, the rocket the V-2. At Peenemünde, factories arose to begin full-scale production.

There was no opportunity to begin building V-2s that year, however. In England, an astute Women's Auxiliary Air Force photo-interpreter spotted the V-2 site on a reconnaissance photo. Further surveillance flights and some intense espionage confirmed what was happening. On August 17, 1943, some 600 aircraft dropped 1,500 tons of high explosive on the site. Approximately 800 people were killed.

No further production work could be carried out there. Instead, the rockets would be made by slave labor at a camp near Nordhausen, 120 miles southwest of Berlin in the Harz Mountains, safe from Allied bombs. Here, starting in late 1943, tens of thousands of prisoners produced about 900 V-2s a month in barbarous conditions that killed some 20,000 of them. Von Braun knew about these conditions, since his brother was a manager in Nordhausen. But he was in charge of a huge team of research scientists at Peenemünde—a community of about 5,000 people, including wives and children—and had no direct responsibility for the V-2 factories themselves.

For a few weeks the V-2 showed some promise of developing into the weapon of terror that would fulfill Hitler's hopes. Suddenly, other Nazi leaders started to take an interest, in particular Heinrich Himmler and senior officers of the SS, Hitler's elite "Protection Squad," and their secret police colleagues, the Gestapo. In February 1944, von Braun himself was briefly arrested by the Gestapo and held for two weeks on the grounds that he wasn't really interested in the V-2 as a weapon. What he wanted, they claimed, was a space rocket (quite true, as it happened, though there was little distinction between the technological advances needed for the two different purposes). Dornberger went straight to Hitler and said that without von Braun there would be no V-2 at all. He was released, and work continued.

A new terror begins

The V-2 assault opened on September 6, 1944, when two rockets aimed at Paris from The Hague misfired. Two days later, shortly after dawn, a V-2 was fired toward London. At 6:40 a.m., Londoners were awakened by a huge blast. The V-2 had blown an immense crater in part of Chiswick. Fortunately, the missile had penetrated so deep that the blast was contained inside the crater, and only three people were killed. Supposedly, von

TERROR WEAPON Probably heading for London or Antwerp, a V-2 takes off from a clearing in the woods in late 1944. The V-2s did not need permanent launch sites, and so were transported to any convenient location within range of their intended targets.

BLITZED AGAIN Rescuers drag survivors from a house in east London that had been hit by one of Hitler's "weapons of retaliation" in 1944. The war was almost over, but a new terror was falling from the skies.

V-2 SPECIFICATIONS

Length:	46 feet
Diameter:	5 feet 5 inches
Weight:	12 tons
Payload:	1 ton
Range:	180-200 miles
Thrust:	24 1/2 tons
Max. trajectory height:	70 miles
Flying time:	5 minutes
Max. speed:	3,500 mph

Warhead

Control compartment
(batteries, guidance
systems)

Alcohol (fuel) tank

Refrigeration, for keeping
the fuel liquid

Alcohol outlet valve

Oxygen tank, with alcohol
delivery pipe passing right
through it

Turbine for compressing the
alcohol and oxygen

Rocket motor
(combustion
chamber)

Rudder vanes

Braun was delighted that the rocket was a success, but was disappointed that "it landed on the wrong planet."

A little later, another blast came, from Epping, England. Since the rocket came literally out of the blue, traveling at over four times the speed of sound—so it could not be heard until after it had arrived—the explosion was a complete mystery to the public. Other mysterious blasts followed, and rumors soon spread that they were caused by a new type of German rocket. The government knew perfectly well what caused them, because a stray V-2 had fallen into British hands when it crashed in Sweden in June 1944; but to avoid spreading panic, officials tried to dismiss the blasts as gas explosions. No one believed this explanation. Only after three months did Churchill reveal the truth.

For a while, there was genuine terror, for a V-2 could flatten 50 four-story buildings. The worst strike occurred at lunchtime on November 25, when a V-2 landed on the Woolworth's store in Deptford, England, killing 168 people.

Within weeks, however, public fear began to dissipate. The V-2s could not be produced or fired in anything like the numbers Hitler wanted. After being manufactured in Nordhausen, each one had to be transported by road or rail, accompanied by some 30 support vehicles, to a launch site within reach of a useful target. This was at a time when Germany was in fast retreat from both east and west, and when anything moving was liable to be attacked by Allied aircraft. The alcohol that fueled the rockets was in desperately short supply. And these were delicate, sophisticated devices that often went wrong: they blew up on the launch pad or before their fuel ran out, their guidance systems malfunctioned, their warheads failed to explode, and they were accurate only to within about 10 miles.

As a result, the V-2's impact on the course of the war was negligible. In seven months, some 4,000 were fired, over half of them targeting Antwerp, which was already in the hands of the advancing Allies. Some 1,100 landed on England, half of them on London. In all, 2,754 people were killed by V-2s, and 6,500 injured. By contrast, Allied air raids on Germany killed tens of thousands in major cities such as Berlin and Dresden.

Yet if the V-2 did not alter the course of the war, this was because its potential had not been realized. Only in the United States, where scientists were well advanced on their own secret atomic weapons program, were there people who could appreciate the true significance of Hitler's V-weapon.

Escape to the West

By January 1945, Allied advances had forced the Germans back so far that even V-2s could not reach useful targets. Von Braun and his staff at Peenemünde were trapped between the Russians closing in from the east and the Americans advancing from the west. Bombarded by conflicting orders to

V-2 BASE In spite of Allied bombing raids, the rocket research station at Peenemünde remained open for work. This aerial photograph from July 1944 shows several rockets ready for firing.

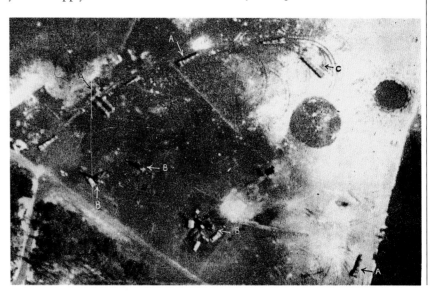

evacuate and to fight to the end, von Braun and his colleagues made a secret decision to surrender to the Americans as soon as they could. Their problem was how to get themselves and their information past military roadblocks and Gestapo checkpoints.

Von Braun picked out the most useful order—one from the Armaments Ministry commanding that rocket research should continue at Bleicherode, near Nordhausen. He then told his staff to pile their 14 tons of documents, which detailed rocket research over 13 years, into cars and railway carriages, and invented a title for their operation—*Vorhaben zur Besonderen Verwendung* ("Project for Special Use")—which they emblazoned on their vehicles. It sounded official, and secret, and it worked.

In February, von Braun and his team headed south, bluffing their way through checkpoints, until in March they reached Bleicherode. They were welcomed by a senior SS officer, General Hans Kammler. In early April, Kammler ordered the transfer of von Braun and some 500 of his top staff, including his brother Magnus and Walter Dornberger, to an army camp in Bavaria, in the south of the country.

Von Braun faced another hard choice: he and his reduced staff would be unable to take all their documents with them. Two members of the team therefore spirited away the 14 tons of files and hid them in a mountain mining tunnel. Meanwhile, the rest of the group, including von Braun and Dornberger, were moved on southward to Oberammergau, close to the Austrian border, where they scattered themselves among the local population.

SEIZING THE MOMENT In April 1945, the U.S. Army captured 100 V-2s on their assembly lines at Nordhausen. The find included motors (below) and whole rockets, so the Americans moved swiftly to remove everything they could by train (right) before having to hand over the site to the advancing Soviet armies.

At this time, in April, the U.S. Army was advancing rapidly toward Berlin when it received orders from Washington to capture as much V-2 hardware as possible from the Nordhausen factory. Whether this mission would be possible remained uncertain for a while, as there was a kind of halfheartedness about American military policy.

In February, the Western Allies and the Soviet Union had agreed at Yalta that a huge area around Berlin—the future East Germany—would be occupied and adminis-

FRIENDLY SURRENDER Wernher von Braun, cheerful despite a broken arm, hands himself and his staff over to Charles Stewart (left) of U.S. Army Intelligence. The man directly to the right of von Braun is his brother, Magnus, who first contacted Stewart in Austria.

tered by the Soviets. And Nordhausen lay in this zone.

As it happened, the United States got there first. When they entered the underground factory on April 11, American troops found an astonishing sight: 100 V-2s still on the assembly lines. Knowing they had little time, the Army troops rushed to secure their find before they were forced out by the incoming Russians, as had been agreed at Yalta.

News of Hitler's death on April 30 indicated to von Braun that the end was in sight. He dispatched Magnus, who spoke English, to make contact with the Americans in Reutte, 50 miles away in Austria.

Charles Stewart, an agent with Army Intelligence, was in Reutte processing Nazi officials, when Magnus von Braun was brought in. "He explained that his brother and some 150 of the top German rocket personnel were lodged in an inn behind the German lines," Stewart recalled later. "They wished to join the Americans to continue their work in rocket development. They had

selected the Americans, as they were favorably disposed to this country generally, and also because this country was the one most able to provide the resources required for interplanetary travel."

There was little time to waste because the SS now had orders to kill German scientists to prevent their information from falling into Allied hands. It took a few days for Stewart to get the ear of his superiors, but when his news got through the clogged channels of military bureaucracy, officials in Washington moved fast to net von Braun, his colleagues, the V-2s and the documents.

The war in Europe ended on May 8. By then, British sources had provided American Intelligence officials with a list of German rocket experts. The two scientists who had

hidden the 14 tons of rocket research papers were traced, and with their help the documents fell into American hands on May 21. The first trainload of V-2s left Nordhausen the following day; the last left on May 31.

In all, 341 trucks carried 100 V-2s to Antwerp. They were then taken by ship to New Orleans, and from there were transported to the New Mexico desert. The head of that operation, Colonel Holger Toftoy, then recommended that the German rocket scientists themselves be brought to the United States. This was achieved in a program, code-named Paperclip, which was thrown open to all German scientists.

Meanwhile, the Soviet army had moved in at Bleicherode and Nordhausen to seize what remained of the Peenemünde group, now reduced to about 3,500 lower-ranking scientists, technicians and their families. The Russians established their own group of German rocket scientists, keeping them in Bleicherode for another year before transporting 200 of them to the Soviet Union.

It was the United States, therefore, that reaped the greater benefit from von Braun's work. And the wisdom of American policy soon became clear when two atomic bombs were dropped on Japan in August 1945.

Hitler had realized that the V-2 needed to deliver more than 1 ton of conventional high explosives to be truly effective. Now, with the discovery of atomic weapons, the V-2 acquired a sudden terrifying potential. If a rocket could be made to carry an atomic warhead, it would have a capacity for destruction that was beyond Hitler's dreams.

AIMING HIGH America was not the only country interested in the V-2. Here a group of British troops examine a rocket with the assistance of captured German experts.

OUT OF THIS WORLD

THE YEARNING FOR SPACE TRAVEL COULD NOT HAVE BEEN TRANSFORMED FROM DREAM TO REALITY WITHOUT HUGE SUMS OF MONEY TO HARNESS AND PUSH FORWARD THE CENTURY'S TECHNOLOGICAL ADVANCES. THE IMPETUS FOR THIS WAS PROVIDED BY TWO BASIC POLITICAL MOTIVES—THE URGE TO DESTROY AND TO PROTECT. THE SECOND WORLD WAR HAD GIVEN HUMANITY ITS FIRST SPACE HARDWARE; THE COLD WAR FORCED THE INVESTMENT NEEDED TO PUT MEN INTO ORBIT.

INTO ORBIT

BOTH THE U.S. AND THE U.S.S.R. BEGAN WORK ON PROGRAMS THAT WOULD PUT THE FIRST MAN-MADE OBJECTS INTO SPACE

By February 1946, more than 100 German scientists were working in the cactus-strewn desolation of Fort Bliss, a military base spanning the border between Texas and New Mexico. The first rocket flight at White Sands Proving Grounds a few miles north in New Mexico would come just two months later.

The team had realized that V-2s could never become sufficiently accurate long-

STAGE TWO American scientists launch a V-2 carrying a smaller WAC-Corporal rocket as an upper stage. The mission was part of an ongoing program designed to research high-speed, high-altitude missile flight.

range missiles, because their guidance systems stopped working when the fuel ran out. They also needed a 1 ton payload in the nose for balance, so several were loaded with lead to provide ballast. Yet over the next six years, the German-American team steadily used up their supply of V-2s. One pointer to future designs came in February 1949, when a V-2 carried aloft a WAC-Corporal, an upper-atmosphere research rocket, also called a "sounding rocket." This successful two-stage ascent set an altitude record of 244 miles.

The following year, when the Korean War broke out, von Braun and his team were transferred from Fort Bliss to Huntsville,

Alabama, where they settled at a nearby army ordnance base, the Redstone Arsenal. From the research work conducted here, the Redstone rocket emerged in 1953. The Redstone was a short-range missile capable of carrying a nuclear warhead 200 miles.

The Viking rocket

In the case of long-range (or high-altitude) rockets, a major advance came with the Navy's introduction of the Viking, the most powerful of two dozen different rockets that could act either as weapons or as sounding rockets. As a weapon, the Viking was designed to be launched from an aircraft-carrier and was intended to carry a 500-pound payload 100 miles.

The V-2 could do this easily, but being built of steel in order to survive the heat of re-entry into the atmosphere, it was vastly over-engineered for the task. When the Viking was used for high-atmosphere research, the rocket itself was jettisoned and only the scientific payload survived, parachuting back to Earth. As a result the Viking could be one-third the weight of a V-2. The Viking also had new steering, avoiding the V-2's wasteful system of exhaust rudders. In the Viking, the motors themselves moved to redirect the rocket, as an outboard motor steers a dinghy.

The first Viking flew in 1949, the last in 1957. In May 1954, Viking number 11 lifted 825 pounds to a height of 158 miles. It also took the first photographs of the Earth from space, including a shot of a hurricane—a pointer to the future use of space cameras for meteorology. The whole project was considered a triumph.

The American authorities were not especially interested at this stage in very long-range rockets as weapons; nor did they care much for the idea of using rockets for space travel. Those who made the decisions and supplied the money could see no purpose in them. The Air Force, which held prime responsibility for long-range strategy, was wedded to bombers, and it wanted cruise missiles (long-range, low-flying, jet-powered missiles) to support them. The B-52 bomber,

1949 Using captured V-2s, the U.S. tests two-stage rockets

1950 William Bollay develops a powerful new motor

1951 Willy Ley holds a symposium on the future of space travel

1952 U.S. explodes the first H-bomb

COUNTDOWN TO LAUNCH Two Army experts check the firing panel of a Redstone rocket, which was developed as a short-range surface-to-surface missile, capable of carrying a nuclear warhead.

inspiration came from the Air Force, which was interested in developing a pilotless, long-range, jet-powered cruise missile. It also began researching, but not developing, an intercontinental ballistic missile, or ICBM.

The work on cruise missiles was done by William Bollay, the design genius of a company called North American, a wartime aircraft manufacturer. Working at North American's test center on a hilltop overlooking the San Fernando valley, just north of Los Angeles, Bollay pointed out that a slow-moving cruise missile presented an easy target. How about a high-flying, rocket-powered missile? Bollay received the necessary approval and got to work on a pro-

ject code-named Navaho. Its result was a rocket with wings, powered by a new motor with more than 8 tons more thrust than that of the V-2.

By March 1950, Bollay's new motor was ready for testing. With von Braun and several others watching, the engine stood almost hidden by fuel tanks and scaffolding among the red rocks of the hilltop site. "Oxygen valve open!" yelled Bollay's supervisor—and the rocket exploded. A designer had used a common form of steel in the engine, unaware that it would become brittle when frozen by liquid oxygen before ignition. A few weeks later, strengthened with stainless steel, the motor worked to perfection.

When it was completed in 1956, Navaho was a 95-foot, 140-ton pilotless rocket-plane

which appeared in the early 1950s, would dominate long-range strategy for 20 years. The Navy also wanted missiles, but only if they were small enough to be fired from ships and submarines. And the Army needed short-range rockets, such as the Redstone, for battlefield use.

These limited and disparate aims were united by an air of complacency. In 1946, the United States had a monopoly on atomic weapons and delivery systems. No one believed that the Soviets would develop matching power and expertise in the foreseeable future. Besides, the concept of an intercontinental rocket was forbidding. The Hiroshima bomb had weighed 5 tons. A rocket powerful enough to carry it would be a good 20 times heavier than the V-2. Its nose cone would burn up on re-entry into the Earth's atmosphere.

And even if the rocket did survive, no guidance system yet existed that could steer it to a distant goal such as Moscow. An American B-36 bomber, on the other hand, could fly to Moscow, reduce it to rubble and return, all without refueling.

Cruise missiles and ICBMs

Nevertheless, research was under way on a number of projects that would eventually help to take men into space. Ironically, the

HIGHEST VIEW YET Viking, an enormous single-stage rocket built by the Navy, roars off on May 24, 1954 (right). It reached an altitude of 158 miles, from where it took the first near-orbit photographs of the Earth, including this one (inset).

PIGGY-BACK ALOFT   A Navaho cruise missile is launched on the back of a booster rocket in 1958. Many of the systems developed for the Navaho—including engines and guidance systems—fed directly into designs for ICBMs.

with rakish delta wings; it was carried aloft by a booster. In the stratosphere, traveling faster than sound, the Navaho's ramjets would kick in, taking it to a cruising height of some 60,000 feet. The Navaho was capable of climbing to 80,000 feet at 1,800 mph; it could pilot itself by checking its position against the stars, and could recognize its target with infrared sensors that picked up the heat of the city it was aimed at. The Navaho flew only 11 times, but from it came a mass of new technology, including engines, guidance systems and fuel, that would eventually lead to the Moon rockets of the 1960s.

At the same time, the Air Force contracted research into a true ICBM. Its mentor was Karel Bossart, of Belgian origin, who was known universally as Charlie. He produced plans for the most powerful rocket yet conceived. Standing 160 feet high and weighing 122 tons, it would be lifted aloft by five of Bollay's massive engines, modified to burn kerosene rather than alcohol, and delivering 375 tons of thrust. Bossart called the rocket Atlas, after the parent company of his own employers, Convair. Its time had not yet come, but it provided experience in designing a multistage rocket with a range of 5,000 miles.

When the United States exploded its first H-bomb in 1952, the problems of guidance and accuracy shrank. The first H-bomb created a fireball 3 miles in diameter, which vaporized an entire island: Eniwatok Atoll in the Pacific. With such explosive power, a bomb need only land within miles, not feet, of its target. Moreover, it did not need to be as big as the first atomic bombs. A smaller payload, delivered with less accuracy: suddenly, an ICBM seemed a real possibility.

Developments in the Soviet Union

Meanwhile, events in the Soviet Union were acquiring a momentum that would drive both superpowers into intense rivalry—and also into space.

In 1928, Stalin had proclaimed his ambition: "We are 50 or 100 years behind the advanced countries. We must make good the lag in ten years. Either we do it or they crush us." With industrial espionage and enforced industrialization, he might have achieved his ambitions, but for the Second World War. After Hiroshima and Nagasaki, Stalin demanded of his officials and scientists: "Provide us with atomic weapons in the shortest possible time." The U.S.S.R. exploded its first atomic device in 1949. For a delivery system, the Russian designer Andrei Tupolev copied an American B-29 bomber that fell into Soviet hands when forced to land at Vladivostok in 1944.

The Soviets also worked on ballistic missiles, but with a different agenda from that of the Americans. By the late 1940s, the Soviet empire reached to the heart of Europe, having swallowed East Germany along with the rest of Eastern Europe. V-2s might not defend Moscow from an American B-29, but they could certainly carry warheads into NATO territory. The Soviets had been left with only the leftovers after the

Americans had stripped Nord-hausen of its V-2s. And only one high-level factory manager had opted to assist the Soviets: Helmut Gröttrup, second in charge of guidance and telemetry.

At this time, Sergei Korolev, who had spent the Second World War under arrest researching fuel systems for aircraft, was working in a prison research group in Omsk under the control of his old friend Valentin Glushko. Glushko was free, and Korolev still technically a prisoner, but the two were sent to see what could be made of the V-2 remnants in East Germany. Soon Nordhausen was put back into commission under Gröttrup. A year later, the Germans and Russians had built 15 new V-2s. Then, in October 1946, Gröttrup and 100 other German technicians were seized and taken to Moscow to work on improving the V-2. They lived and worked in appalling conditions for years while their brains were picked clean of information. Only then were they allowed to go home.

Under the auspices of Korolev, who had been set free, the Soviet Union began developing bigger rockets. Adapting the V-2, Korolev produced his R-2, with a range of 375 miles, firing it from bare and desolate steppe-land near Kapustin Yar, 75 miles east of Stalingrad. After the Soviet Union exploded its first atomic bomb in 1949, Korolev responded to Stalin's demands for a delivery system by proposing his R-3, with a range of 1,800 miles, though he had no idea how to build engines powerful enough to carry it.

The breakthrough came in 1952. Until then, more power had meant more engines, with a corresponding increase in the risk of failure. The turbo-pumps, in particular, were a

SOVIET SUPER BOOSTER Korolev's massive R-7, with four boosters encircling a core engine, was the workhorse that would put the world's first satellite, Sputnik, into orbit in October 1957.

SCIENCE FICTION IN THE 1950S

SOVIET-AMERICAN RIVALRY, ADVANCES IN SPACE SCIENCE AND WIDESPREAD GULLIBILITY ENCOURAGED A NEW TYPE OF FICTION TO BLOSSOM

Science fiction, originally little more than stories about "bug-eyed monsters" and fantasy dressed as science, came into its own in the two decades after the Second World War. Critics explain this boom as a conjunction of several influences. The existence of two competing superpowers and two world views stimulated an interest in those living under different systems. "Alien" might easily stand for "communist" or, in a world increasingly dominated by companies and governments, something more sinister and closer to home. Science in general made technological wonders imminent. In particular, the growth in rocketry made space travel seem inevitable. Finally, astronomy was not yet advanced enough to destroy the mystery of other worlds. In the 1950s it was still possible to write about aliens on Mars and Venus, without seeming eccentric.

These factors resulted in the growth of an entirely new literary genre. In the United States and Britain in the 1950s, some 200 science-fiction novels and short-story collections were published annually, while half a million subscribers devoured about 20 magazines every month. American publications dominated, led by *Astounding Science Fiction*, *Amazing Science Fiction* and *Galaxy*. In Britain, a generation of schoolboys avidly read the *Eagle* comic to follow the adventures of Dan Dare, Pilot of the Future, in his everlasting battle against his sworn enemy, the bubble-headed, green, Venusian genius, the Mekon.

In mainstream science fiction, the writing, mainly by men, was notable for its imagination and expertise. John Cambell, editor of *Astounding*, insisted that the science in his fiction be well worked out, and attracted writers such as Isaac Asimov and Robert Heinlein. Many of the subjects examined were doom-laden. Humanity would be threatened with extinction (a growing fear, given the increasing power of nuclear weapons) for any number of reasons: from natural disasters, alien invasion, or for reasons beyond the power of mere mortals to fathom. People were often the victims of unknown forces, particularly technology run wild. Robots and computers attracted writers' attention as human inventions that might one day control humankind. Time travel was another popular theme, while many writers examined the sociological effects of technological innovation, as Fred Pohl and C.M. Kornbluth did in *Gravy Planet*, a portrait of the world dominated by advertising.

One constant preoccupation, given the increased certainty of manned space travel, was the idea of contact with alien civilizations. Previously, this had been the domain of horror writing, but now it warranted serious treatment. Perhaps the most notable exploration of the theme was in Stanley Kubrick's film *2001: A Space Odyssey* (1968), based on the proposition that human civilization developed from the discovery of an alien artifact, and the suggestion that space travel would lead to further contact, and another leap forward.

DAWNING REALISM   Stanley Kubrick's film *2001: A Space Odyssey* developed three major themes of maturing science fiction: the realities of space travel, contact with aliens, and the challenge and implications of creating artificial intelligence.

ALIENS GALORE   Many sci-fi magazines of the 1950s were little more than "space Westerns," with BEMs—bug-eyed monsters—as the "baddies," and spacemen as the "goodies." Heroines were notoriously well-endowed.

nightmare since each engine required its own pump. Glushko suggested a way forward: what if one giant turbo-pump serviced all the engines?

At this moment, war in Korea revealed a flaw in Soviet strategy. American B-29s proved vulnerable to new Russian-built jet fighters. This meant that the Tu-4, Tupolev's version of the B-29, was also outmoded. Then, in November 1952, the United States exploded its first H-bomb. Stalin responded by giving orders for the creation of an ICBM that could carry nuclear weapons.

Korolev's giant R-7

The concept that Korolev then designed—a development of his R-3 combined with Glushko's engines—was stunning in its power and originality. At its heart would be four chambers firing together to produce some 75 tons of thrust. Surrounding the core would be four strap-on boosters, providing a total lift of 392 tons—nine times the power of any other Soviet rocket, and more than the American Atlas. It would be powerful enough to carry a 5-ton thermonuclear warhead some 5,000 miles, far enough to reach the United States. This dream vehicle was designated R-7.

Korolev's re-emergence from obscurity to become the driving force in Soviet rocketry was a tribute to the strength of his intellect and personality. He had emerged from his years of persecution with his messianic determination untouched. It shone out from his features: high forehead, fierce black eyebrows, expressive and deep-set brown eyes.

He was formally rehabilitated only after Stalin's death in 1953, when he was elected to the Soviet Academy of Sciences. Even then, however, he remained a shadowy figure, being referred to simply as "the chief designer," and publishing under the pseudonym "Sergeyev." Like von Braun, his inspiration was to get into space. And like von Braun, the only

DUAL PURPOSE The Atlas rocket, designed initially as an ICBM, was also powerful enough to launch numerous early space missions for America.

way he could get the funds was to make missiles, which he did with equal passion.

Korolev certainly impressed Stalin's successor, Nikita Khrushchev, who, like other leading Party members, had been kept in the dark about Soviet rockets. Khrushchev later wrote: "I don't want to exaggerate, but I'd say we gawked at what he showed us as if we were sheep seeing a new gate for the first time. When he showed us one of his rockets, we thought it looked like nothing but a huge cigar-shaped tube, and we didn't believe it would fly. Korolev took us on a tour of the launching pad and tried to explain how a rocket worked. We were like peasants in a marketplace. We walked around the rocket, touching it, tapping it to see if it was sturdy enough—we did every-

thing but lick it to see how it tasted."

Korolev was soon to benefit from the good impression he had made. A major problem in the development of ICBMs was the weight of nuclear warheads. Whereas the United States decided to wait for smaller warheads before developing a rocket to deliver them, the Soviets simply opted for the massive five-engined, 100 ton R-7.

The Atlas rocket

In 1953, American leaders were becoming obsessed with what they saw as Soviet aggression. In hindsight, such beliefs seem an overreaction. But at the time, officials were weighed down by the certainty that if they did not respond, then, in the words of presidential adviser Paul Nitze, "our very

DREAMING OF A SPACE STATION In 1955, Willy Ley (left) and Wernher von Braun demonstrate their ideas for a three-stage rocket with the power to place a space station in orbit around the Earth.

independence as a nation may be at stake."

In 1954, a top-level committee, the Strategic Missiles Evaluation Group, recommended that the Atlas rocket be produced as the new key component in national defense. After tense lobbying, the armed forces and the Pentagon agreed. The contract went to two committee members, Simon Ramo and Dean Woolridge, who left Hughes Aircraft to set up their own corporation. This was a project of unrivaled complexity, beyond the capacity of any one company. Answering to the Air Force's Brigadier General Bernard Schriever, Ramo-Woolridge eventually ended up at the center of a web of 220 companies, representing the cream of the American aviation and electronics industries.

There were few precedents. As problems emerged, ideas, experiments, new techniques and whole new industries were born, many of them interrelated. The guidance

system was one example. It would have at its heart a gyroscope, a heavy wheel spinning so that its mountings would be sensitive to any movement. But such sensitivity meant that the whole device had to be encased in fluid. In the manufacturing process, no speck of dust could be allowed. The air was changed every nine minutes, under pressure so that any leak went outward. No one was allowed to tear paper or use erasers. In use, the gyro could be made to send signals through a computer. But no computer was light and sturdy enough to be included in the rocket. Ground-based computers would be required, and they would have to communicate with the rocket by radio. That meant reliance on ground installations, which were vulnerable and limited in range.

Another problem was that of re-entry. The warhead, dropping from inner space, would hit the atmosphere at 16,000 mph, generating enough heat to turn normal rock or metal into a meteor. Common sense dictated that the nose should be needle-sharp, like that of a jet fighter. In fact, research showed exactly the opposite. A blunt nose would compress air in front of it, which

would act as an insulating blanket. In the end, designers joined the two concepts, opting for a blunt nose cone enclosed by a shell of reinforced plastic that would be allowed to burn away, meteor-fashion, losing heat as it lost bulk.

As contracts were handed out, Schriever opted for a management process known as "parallel development," by which two contractors worked on the same problem. It was an insurance policy. If one failed, the other could continue. And if both worked, their systems would be interchangeable. Such an approach had another benefit: it could sustain a second ICBM program—Titan.

Neck and neck, the two superpowers raced to be first to create the machines that could secure, or end, civilization.

Help from the media

All these developments, many of them secret, had a very public effect. In the early 1950s, the technology was in place to turn dreams of space travel into reality. There was still no greater dreamer, and no greater realist, than Wernher von Braun.

In 1951, the science writer Willy Ley set up a symposium on the future of space travel that was covered, among many others, by Cornelius Ryan, author of *The Longest Day* and associate editor of *Collier's* magazine, which then had a readership of 10 million. At the symposium, Ryan met von Braun, who took the opportunity to sell his dreams: manned space rockets, a space station, cargo rockets weighing 7,000 tons, a Moon mission by the mid-1970s. Later, next century, there could be a Mars mission.

Ryan took up the idea. In 1952, *Collier's* ran a series of beautifully illustrated articles on the subject of space travel. It was to be more than an epic for humankind; it had a defensive purpose. One problem shared by both superpowers was the need for information about what the other side was up to. What was needed was a spy-in-the-sky: an Earth satellite that could focus a camera on any part of an opponent's territory, and drop bombs if necessary. *Collier's* delivered an urgent warning, "that the United States must immediately embark on a long-range development program to secure for the West 'space superiority.'"

The articles unleashed a media frenzy. By the middle of the decade, all Americans knew their country was on the verge of acquiring the technology for space travel, taking the lead in a great new adventure for all humankind, and at the same time asserting the supremacy of American ideals and scientific know-how.

Much research and thinking had already gone into satellites, from which it was clear that they would open up a new era of scientific research and military observation. Further support came in the form of an international scientific council, which declared 1957-8 International Geophysical Year (IGY) to coincide with a time of maximum sunspot cycles. In 1954, IGY scientists agreed to push for a satellite launch to assist them in the study of sunspots. President Eisenhower gave his backing to the idea the following year.

As it happened, science and military necessity made perfect partners. The Air Force, in the person of General Curtis Le May, head of Strategic Air Command, was desperate for information. Le May's bomber crews needed to know more about their targets and the state of Soviet air defenses. Until then, bombers had managed to penetrate the leaky Soviet radar system, but with less than total reliability. The Soviet Union's H-bomb had been a nasty surprise, and no one knew about the state of their long-range missile program. The program's command structure was also unknown, as was Korolev's very name.

Out of the blue, on May Day, 1954, in the great Red Square parade, the Soviets unveiled a new jet bomber called the Bison. A week later, one of Le May's B-47s, on a mission over Soviet territory, was damaged by a MiG-17 interceptor. Clearly, the Soviets were catching up, and improving their defenses. A new American spy plane, the high-flying U-2, would help, but only until the Soviets developed better radar systems. "We have an offensive advantage," concluded a American defense report in February 1955, "but are vulnerable to surprise attack."

The Air Force moved fast to solve the main outstanding problem: how to send back photographic images from space. One possible answer was to use radio. Another was to eject exposed film in a capsule that would parachute to the ground; the latter scheme was selected because it guaranteed higher quality images. Both suggestions undermined von Braun's belief that only astronauts could do what was required. There would be no

SLEEK BUT UNSTABLE The Navy's Vanguard was expected to carry a satellite into orbit. It did not. Several, like this one at Cape Canaveral in June 1957, exploded on takeoff.

FUTURE VISION The American public was gripped by the idea of space travel, as presented by *Collier's* magazine. Here, a three-stage, manned space rocket lifts off.

space station, at least not yet. A satellite would do the job.

There remained the issue of how to dress up a surveillance program in scientific clothing. A von Braun rocket, produced in an Army base, would seem deliberately provocative. To the Army's chagrin, Defense Secretary Charles Wilson banned the Army from developing long-range rockets (though he did not specifically ban space research, a loophole that von Braun immediately seized upon). Instead, Wilson opted for the Naval Research Laboratory as a suitable organization to back an IGY satellite.

Rival American rockets

But which rocket would the IGY satellite fly in? There were several possibilities. The army was almost ready for a satellite launch with a new three-stage rocket from von Braun's stable: a modified Redstone called the Jupiter-C, which in September 1956 actually flew higher, faster and farther than any other rocket, simply in order to test nose cones. The Navy, which co-sponsored the Jupiter, also had the Vanguard, while the Air Force had three missiles in development: Atlas, Titan and Thor. Inter-service and inter-company rivalry was intense, and it was sometimes hard for rocketeers to remember that the enemy was supposed to be the Soviets, not the guys down the road.

Work went forward at a frantic pace, principally at the Redstone Arsenal in Alabama, where von Braun's Jupiter took shape, and at the main test site: the desert base of Santa Susannah—called "the Hill"—where engineers ran a futuristic array of buildings, tanks and testing equipment. Tucked away in ravines were massive steel structures built into solid rock to hold rocket motors. In the words of *Time* magazine, when a rocket motor was being tested: "A sound like the rumble of doomsday rolls around the rocks, making the flesh quiver like shaken jelly."

Meanwhile, a launch facility was being developed at Cape Canaveral on the Florida coast. The Pentagon had picked the site as a missile test center in 1947 because it had a clear range eastward into the Atlantic. The low-lying spits, separated from the mainland by the Banana and Indian rivers, were swampy, their salt marshes infested with mosquitoes. The nearest town, Titusville, was 30 miles away, with 2,604 people. But in 1950, that had all begun to change. Bulldozers gouged roads from the swamps. Motels sprang up. Launch pads ranged along the coast, with a scattering of administrative blocks and assembly buildings. It was from here that most of the test rockets were fired.

Coordinating a test launch was said to be like conducting an orchestra. The flight director faced a row of controllers, each with a console showing the state of his system. The countdown took 400 minutes, but it could last indefinitely, because if there was a problem the count went on hold until the trouble was fixed. Some counts lasted 30 hours. At the very end of the count, each controller would give the thumbs-up. When everything was "go," the test coordinator pushed a button to fire the engine. As often as not, a test ended in disaster, and both the successes and the explosive failures happened in full public view. There was even a ladies' luncheon club, the Missile Misses,

whose members gathered to watch launches from the beach.

In November 1956 came a decision from Defense Secretary Wilson that the Air Force would handle long-range missiles. The Navy then decided to withdraw from von Braun's Jupiter project and concentrate on its own new rocket, the Polaris. With Jupiter forced out of the running, the Air Force pushed ahead hectically with its Thor rocket. The rocket was finally ready for launching at Cape Canaveral in January 1957.

Thor lifted a few inches, fell back and exploded. Its liquid-oxygen fuel had been contaminated by a few grains of sand. The second Thor flew for half a minute before it, too, exploded. A safety officer had wrongly thought it was heading for the city of Orlando and blew it up. The third exploded five minutes before launch. Some of those watching wept in frustration.

Von Braun was also showing what he could achieve. In March, a Jupiter exploded after flying 60 miles. Then, in May, one Jupiter flew 1,600 miles, while another carried two monkeys 300 miles high and 1,600 miles downrange.

Toward Sputnik

Meanwhile, in the Soviet Union, Korolev's huge R-7 was taking shape. He had a new cosmodrome, called Tyuratam, at a remote desert railhead 100 miles east of the Aral

PROMISING BEGINNINGS In 1959, this Thor rocket lifted off well from Vandenberg Air Force Base, California, in an attempt to place a satellite in polar orbit. However, the rocket's second stage failed.

THE LAUNCH OF SPUTNIK

Yaroslav Golovanov, Korolev's biographer, describes the launch of the first man-made Earth satellite, on October 4, 1957:

"Korolev arrived early, and left his car on the concrete apron. The wind was cold and piercing. He raised the collar of his heavy old overcoat. From behind him came the wooden voice of the public address system:

"'Attention! Time check in one minute. Prepare to fuel.'

"The liquid oxygen was fuming like the white steam of a locomotive. The rocket was becoming shrouded. Rime was creeping up from the bottom of the oxygen tank, and soon it would be all white. Beautiful.

"Later, immediately before blast-off, he sat with his shoulders hunched on his usual seat at his 'personal' periscope in the bunker. He was shivering slightly.

"'Zero minus one minute! Repeat: zero minus one minute!'

"'Switch to start!'

"Korolev brought his face close to the periscope and felt the unpleasant sensation of cold sweat from his face on the black rubber shield of the eye-pieces. The white oxygen clouds disappeared; the vent valves had been closed.

"'Auxiliary engines pressurized!'

"'Main engines pressurized!'

"'Launch begun!'

"'Ground—disengage!'

"Korolev clung to the periscope. The rocket was right before his eyes. He saw how the cable tower swung away after the command. Now nothing connected the rocket with the launcher.

"'Ignition!'

"'Preliminary boost!'

"He saw an instantaneous flash, a short flicker, before the brown cloud of dust and smoke raging in the whirlwind of the engines rapidly engulfed everything around. A blinding ball of light flared beneath it.

"'Main boost!'

"The rocket hung motionless. It would be a few moments before it rose. It really looked as if it were pondering for a second whether to start or not. How tedious and long were these instants of immobility!

"'Take-off!'

"'There it goes! There it goes!'

"The giant white dagger was racing upwards, its body looking transparent and unreal in the dazzle. Korolev's fingers tightened around the black grips of the periscope, the whole of his thick-set heavy body stiffening.

"Only now did a jubilant voice penetrate his consciousness, gabbling over and over in its excitement: 'All systems stable! Flight proceeding normally! Pressure in chambers normal! All systems stable!' And at last: 'Separation complete!' ...

"A wave of unendurable warmth and gratitude to all the people swept over him. He felt a lump in his throat. 'Is this really all?' [he thought] 'Have we really done it? Of course, of course! It's time to phone Moscow and report. But let it complete one orbit, then we'll report.'"

Sea, in what is now Kazakhstan. From here, rockets could be fired more than 3,000 miles eastward and northward across the wastes of Siberia.

Engine trials began in 1955. The full cluster of engines—four boosters surrounding a central core—arrived at Tyuratam for testing in the winter of 1956. The facilities there were far more elaborate than those at Cape Canaveral, as befitted a rocket with $2^{1}/_{2}$ times the power of America's biggest. R-7 would be assembled on its side, taken out by rail and then lifted onto its launch pad, known as Tyulpan ("Tulip"), a petal-shaped structure of steel girders that would support the rocket and also provide access for technicians. During liftoff, the petals would unfold and release the missile.

Although the R-7's primary purpose was to carry a nuclear warhead, the Academy of Sciences received permission for a program of scientific research. The question was what to send into space. A true scientific satellite would take time to build, which might allow the United States to win the race. What if they put up just a simple, light device?

Korolev himself designed a globe holding nothing more than a few instruments for measuring the density and temperature of the atmosphere, and the concentration of electrons in the ionosphere. Also on board were some batteries and a radio that transmitted a clear beep. It was called Sputnik ("Companion"), or more fully, Prosteishy ("Preliminary") Sputnik.

On May 15, 1957, the R-7 flew briefly. Then a propellant line broke, starting a fire that caused the rocket to explode just $1^{1}/_{2}$ minutes into the flight. In June, a second R-7 failed to ignite—someone, it turned out, had installed a fuel valve upside down. Korolev tried to blame everyone but himself. But his bosses saw through his posturing. "So much stink about what might have been caused by others," said one when he read Korolev's self-defensive report. "So much perfume for your own excrement."

In July, the R-7 flew again, briefly, crashing when the guidance system failed. But in August, everything worked. An R-7 flew 4,000 miles, its warhead blowing up over

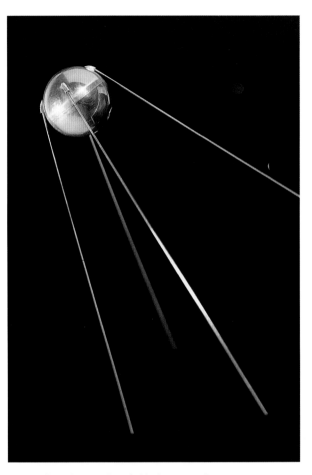

SPACE INSECT Dwarfed by its own radio antennae, the first Sputnik was tiny by modern standards. It weighed only 183 pounds and orbited the Earth every 96 minutes until early 1958, when it re-entered the atmosphere and burned up.

Kamchatka on the Pacific coast. In September, another test flight went well. The R-7 was cleared for an attempt to place Sputnik in orbit.

Sputnik lifted off after sunset on October 4, 1957, its five huge engines firing in unison, propelling Sputnik toward space. The boosters peeled away, leaving the core engines to carry the rocket to 18,000 mph. After five minutes, the rocket was in an arcing path in which its forward speed balanced the Earth's gravitational pull. From the tracking station came a welcome sound, played over the base's speakers: the little *beep-beep*, which proved that Sputnik had survived and was in orbit.

Everyone cheered. Korolev made an impromptu speech: "Today the dreams of the best sons of mankind have come true," he said. "The assault on space has begun."

THE COLD WAR

**THE DREAM OF SPACE TRAVEL WAS PEACEFUL AND IDEALISTIC. YET IT WAS
THE FEAR OF WAR THAT TURNED IT INTO REALITY**

In a magisterial speech in Fulton, Missouri, in 1946, Winston Churchill used two phrases that came to define postwar history. Referring to the empire created by his wartime ally, the Soviet leader Joseph Stalin, he said that an "iron curtain" had descended across Europe. Furthermore, peace had come, but with another type of war: the Cold War.

The Cold War expressed the rivalry between the two superpowers, the Soviet Union and the United States, and inspired their vast expenditure on rocket research. Over three decades, this research would lead to new discoveries, create whole new industries, and send humankind first into orbit, and then into deep space.

The Second World War ended with each of the two superpowers dominant in its own hemisphere. American might was founded on economic strength and a belief in democracy, Soviet might on political control and a vast army. Both sides sought to assert their own interests in any way possible. Allies were wooed and coerced, internal opposition suppressed, ever more destructive nuclear weapons developed, ever bigger rockets built to deliver them.

America's initial faith in a new world order based on cooperation was soon shattered by Soviet intransigence and the growing division of Europe. A communist-backed insurgence in Greece persuaded President Harry Truman to proclaim a policy—which became known as the Truman Doctrine—of checking Soviet ambitions by committing 20 percent of the gross national product to military purposes. A vast aid program, the Marshall Plan, welded western Europe to the American economy. In 1949, the Western nations formed a military pact, the North Atlantic Treaty Organization (NATO); the so-called Eastern Bloc followed suit with the Warsaw Pact a few years later. The eastern and western zones of occupied Germany hardened into two separate countries.

Cold War rivalries dominated other regions of the world as well. In China, Mao Tse-tung led communists to victory in 1949, while

BACK FROM THE EDGE   The Cuban Missile Crisis of 1962 took the U.S. and the U.S.S.R. to the very brink of war. Here an American destroyer escorts a Soviet freighter and its cargo of nuclear missiles away from Cuba.

communist insurgencies in Malaya, the Philippines, Burma and Vietnam gave the impression to American strategists that the world was indeed dividing into two vast and opposing systems. This view received confirmation when Communist North Korea attacked American-backed South Korea in an attempt to reunify a nation divided in 1945. A counter-invasion by American forces brought China into the conflict, forcing the Americans back, until negotiations brought an uneasy peace in 1953.

Stalin's death in the same year eased tensions briefly, but they soon re-emerged with renewed force. Egypt welcomed Soviet aid and repelled a disastrous Anglo-French attempt to occupy the Suez Canal zone. In eastern Europe, an uprising in Hungary in 1956 brought a brutal Soviet invasion. In 1961, the Soviets closed off East Germany and East Berlin behind a wall of concrete and wire. Stalin's successors invested heavily in long-range missiles and

MISSILES, NOT CIGARS   This photograph from a American spy plane reveals a Soviet nuclear missile base being built in Cuba in October 1962. The U.S. Navy blockaded Cuba, and looked set for a showdown with the Soviet fleet, until the latter suddenly turned back.

space hardware. In the early 1960s, Cuba's Communist government under Fidel Castro opened its doors to the U.S.S.R., thus raising the possibility of Soviet rockets being placed within a few hundred miles of Washington. In 1961, President Kennedy was humiliated by a disastrous attempt to land American-backed insurgents on Cuba, at the Bay of Pigs. The following year, the Cuban Missile Crisis brought the two superpowers to within the blink of an eye of full nuclear war. Meanwhile, the looming collapse of non-Communist South Vietnam inspired the U.S. to go to that country's aid. From 1964, America sank ever deeper into a military and political quagmire in Vietnam, which polarized American society.

Gradually, both sides in the Cold War fragmented. China had already broken with the Soviets in 1960. Under General de Gaulle, France withdrew from NATO in 1966. In Europe, communist nations began to work toward greater freedom, though progress in this regard was not always guaranteed. By the end of the 1960s, nations on both sides had more interest in playing down Cold War rivalries than exacerbating them. In the 1970s, the Cold War steadily became a thing of the past, and the space programs inspired by it had to seek economic rather than political justification.

EYEBALL TO EYEBALL As the Berlin Wall goes up in 1961, American and Soviet tanks face each other across the Wall's main crossing point, Checkpoint Charlie (right). Below: West Berliners protest at the sudden imposition of restrictions on travel into East Berlin.

FOOTHOLDS IN SPACE

THE SOVIET UNION'S SUCCESS WITH SPUTNIK, THE FIRST SATELLITE, IGNITED THE RACE FOR DOMINANCE IN EARTH ORBIT—AND BEYOND

On the evening of October 4, 1957, Wernher von Braun and his boss, Major General John Medaris, were in Huntsville, Alabama, attending a party for Neil McElroy, newly nominated to replace Charles Wilson as Defense Secretary. Von Braun, a social charmer and political opportunist as well as a rocket genius, was eager to promote his Jupiter rocket as the nation's best space vehicle. The moment could hardly have been better.

VON BRAUN'S BOSS As chief of the Army Ballistic Missile Agency, Major General John Medaris (left) headed the program that sent the first American satellite into space.

Medaris's public affairs officer, Gordon Harris, dashed in and interrupted: "Dr. von Braun ... They've done it!"

"Done what?"

"The Russians. It's just been announced over the radio that the Russians have put up a successful satellite!"

Sputnik, a 183-pound globe, and the first man-made object in orbit, was following an elliptical path that took it more than 500 miles from Earth.

"We knew they were going to do it!" said von Braun, turning to McElroy in a cold fury of frustration. "They kept telling us, and we knew it, and I'll tell you something else, Mr. Secretary. You know we're counting on Vanguard [the navy's rocket]. The President is counting on Vanguard. I'm telling you right now Vanguard will never make it."

"Doctor," McElroy protested, "I'm not yet the new Secretary. I don't have the authority to ... "

"But you will be. And when you have the authority, for God's sake turn us loose! The hardware is ready! Just give us the green light, Mr. Secretary. Just give us the green light. We can put up a satellite in 60 days."

Medaris added a cautious rider: "No, Wernher, 90 days."

McElroy made no commitment, but Medaris and von Braun assumed he would, especially after the next morning's banner headlines. The name Sputnik at once entered the English language.

President Eisenhower did his best to play down the Soviet publicity coup. All they had done, he told the Press, was to put a "small ball" into the air. But there was no disguising the apparent danger or the possible consequences. A decade earlier, the Soviet Union had been a shattered and backward economy. Now it apparently had rockets powerful enough to carry nuclear weapons to the U.S., to ensure worldwide coverage for its propaganda, and also to open a new era in human history.

"Artificial Earth satellites will pave the way for space travel," crowed the official Soviet press release, concluding that this was proof that socialism worked. It was the propaganda coup of all coups.

Soviet leaders eagerly claimed that success in space equated with success on Earth.

To politicians, Sputnik 1 suggested something even more serious: the possibility that totalitarianism worked. Hitler had made it work temporarily, with terrible consequences. Stalin and his successor had done the same. Perhaps the consequences would be even worse: world war, with nuclear weapons. Even without war, the international impact would be galling. The non-aligned nations of Africa, Asia and South America might take note and choose the Soviet way. Sputnik 1 opened a floodgate of self-analysis. Suddenly "the American way," democracy and materialism, seemed to have bred something effete and debilitating. In the aftermath of Sputnik 1, something close to hysteria swept across the United States.

"Launch a dog by the holidays"

In hindsight, there was no need for such a fuss. Korolev's rocket was a cumbersome and inefficient thing. It took so long to put together that it was highly vulnerable. It needed ground guidance from stations that

SHOWING THE WORLD At the Brussels World Fair in 1958, a visitor peers into a model of Sputnik 2, another space venture that seemed to display the U.S.S.R.'s superiority.

SPACE DOG Laika, the first living creature to be sent into Earth orbit, survived in Sputnik 2 for a week—long enough to prove that life was possible in space.

out. Her body then burned up on re-entry after five months and 2,370 orbits.

Laika's fate inspired a good deal of sick humor in both the United States and the U.S.S.R. Soviet-American rivalry had become "a dog fight," it was said. "What's the difference between Sputnik and Moscow?" Russians asked. Answer: "Up there, the dog's life ends." Animal lovers expressed outrage, some marching in front of the United Nations' headquarters in New York bearing placards of protest.

Bad jokes aside, Laika provided the first real evidence that living things could endure in space. She also fueled additional panic on Capitol Hill because she and her capsule weighed 1,120 pounds, more than six times heavier than Sputnik 1. Democratic leader and future president Lyndon B. Johnson attributed almost magical powers to space technology: "Control of space means control of the world. From space the masters of infinity would have the power to control the Earth's weather, to cause drought and flood, to change the tides and raise the levels of the sea, to divert the Gulf Stream and change temperate climates to frigid."

America's "Flopnik"

America's first attempt at a comeback, with a Vanguard, was an ignominious and highly public disaster. Compared with Korolev's monster, the Vanguard was diminutive, with only 3 percent of the R-7's power, topped by a grapefruit-sized satellite. At its launch, the rocket hesitated, quivered, then toppled slowly over, dissolving into a fireball. The tiny satellite was blown off and rolled away into the surrounding palmetto and scrub, transmitting a noise that sounded piteous to those covering the launch. "Why doesn't somebody go out there, and find it, and kill it?" asked one columnist. Headline writers had a field day: "Oh What a Flopnik!" blared London's *Daily Herald*.

In fact, the United States was well situated for space travel. It had nine ballistic-missile programs under development, with an ever-growing mass of names and acronyms. Even the Vanguard disaster was caused by a minor problem with fuel injection, which was easily fixed.

It was von Braun, the German genius, who salvaged the nation's self-respect. Funded by the Army, which was forbidden to engage in intercontinental missile

were equally vulnerable. And it could only just reach the United States. In the end, its effects were counterproductive. Like the Japanese raid on Pearl Harbor, the shock was so intense that it virtually guaranteed the United States would outstrip its rival.

For the time being, though, America was humbled and Khrushchev was delighted. Making him briefly unassailable in the Kremlin, Sputnik gave Khrushchev enough political clout to continue with military reforms. Naturally, he was eager for more. Next would be the first living creature in space. In Korolev's words, Khrushchev ordered him to "launch a dog by the holidays": in time for the 40th anniversary of the Russian Revolution, in November 1957.

Dogs had been used in high-altitude tests for years, and Korolev had both a suitable dog and a "space kennel." The dog, Laika, as she was called, went aloft on November 3, to a triumphant, but unhappy, end. She lived for about a week, until her oxygen supply ran

1960 U.S. Pioneer 5 launched into deepest space
U.S. TIROS 1 weather satellite launched
U.S. spy satellite Discoverer 14 photographs the Soviet Union

Nose cone
temperature
probe

Low-power transmitter

Cosmic ray and
micrometeorite
instruments

Turnstile
antenna
wire

High-power transmitter

Internal temperature
gauge

Micrometeorite erosion
gauges (located behind
ring)

AMERICAN SATELLITE SUCCESS Scientists of
the Jet Propulsion Laboratory, California,
gather around their brain child, the satellite
Explorer 1 (top), also shown above as a cross-
section. Explorer was blasted into orbit from
Cape Canaveral in January 1958 (left) as the
fourth stage of a Juno rocket—essentially a
modified Jupiter-C designed by von Braun.

research, von Braun had been free to work on his Jupiter rocket, which was not an ICBM but an IRBM: an intermediate-range ballistic missile. It was enough. He modified his three-stage Jupiter-C into a four-stage rocket, renaming it Juno. In January 1958, Juno placed the first U.S. satellite, the 31-pound Explorer 1, into orbit.

Von Braun was not at the launch. He had been ordered to the Pentagon for a press conference, on the assumption that the launch would go well. No confirmation would come until the rocket passed over a tracking station in California an hour-and-a-half after takeoff, after making one complete orbit. The time came, the minutes ticked by. Von Braun paced in tense silence, wondering what could have happened. After 8 minutes, someone shouted from the telephone: "They hear her, Wernher! They hear her!" The satellite was in a slightly higher orbit than planned—hence the delay.

Explorer was lightweight, but contained some heavyweight scientific gear, including a Geiger counter designed by James Van Allen. These instruments recorded important information the Russians had missed: the Earth is surrounded by an intense band of radiation emitted by the Sun and trapped by the Earth's magnetic field. (Actually, Sputnik 2 had recorded the belts, at the high point of its trajectory, but it was then over the Southern Hemisphere, where its radio signal was monitored only by Australian scientists. The secretive Soviets had made no provision for the data to be passed on.)

Shooting at the Moon

Nineteen fifty-eight was a year of frantic action, for both sides were already looking beyond Earth orbit. Behind the scenes in the United States, research was already well under way into a much larger rocket. Though banned from working on a new long-range rocket, von Braun had been exploring ways of using a cluster of eight Jupiter boosters to produce 670 tons of thrust. At first he called it Super-Jupiter, then renamed it Saturn, after the next planet in the Solar System. To supply the immense amounts of fuel needed by such a rocket,

von Braun proposed a cluster of tanks. In late 1958, Rocketdyne received a contract to upgrade Jupiter engines, and in early 1959, Saturn went into full-scale development.

At the same time, elsewhere, researchers were looking into the use of liquid hydrogen instead of alcohol, kerosene and other fuels. The main problem was that liquid hydrogen needs to be kept even colder than liquid oxygen. But it delivers greater power than the other fuels, and it is safer because it dissipates when released, rather than forming an explosive cloud near the ground. In mid-1958, the Air Force was ordered to develop a hydrogen engine for a new upper-stage rocket called Centaur.

Meanwhile, Korolev was laying the groundwork for more immediate, short-term triumphs. Always aware of the need to give bigger rockets a military justification, he suggested that nuclear weapons might be dropped from orbit. To get them up there, he would need a new upper stage to carry the payload, which could initially be made up of scientific instruments. As a test for accuracy and reliability, he proposed to shoot the rocket directly at the Moon. Success would provide Khrushchev with a military advantage, scientific research, space exploration and terrific propaganda all in one.

Approval came quickly. "Comrades," Korolev told his staff, "we've received an order from the government to deliver the Soviet coat of arms to the Moon." He pushed himself like a man possessed, and his staff too. He flew back and forth from Moscow at night, sleeping on the plane, in order not to waste a working day.

At the same time, one line of research in the United States also involved the development of a rocket that would have the capacity to shoot for the Moon. Originally designed to test a nose cone for Atlas, the rocket would be a Thor first stage with a Vanguard second stage, a combination known as Thor-Able. Roy Johnson, head of a new rocket research department in the Pentagon, initiated "Project Red Socks," a mission to put a satellite in lunar orbit that summer. This involved a formidable workload. The contractors, Ramo-Woolridge, had to build a three-stage rocket, design a lunar spacecraft, set up a worldwide network of ground stations to track the craft—and do almost all of it with slide rules. (In 1960, there were only 3,000 computers in the whole world.)

Testing times

Throughout 1958 and 1959, there was a flurry of test launches, explosions and successes, as the two superpowers vied to outdo each

TENSE MOMENTS Anxiety grips controllers at the Air Force Missile Center, Florida, during the countdown to one of the many rocket launches attempted during 1958.

RELIABLE ATLAS An Atlas rocket takes off from Cape Canaveral. Originally designed as an ICBM, the Atlas was adapted to carry satellites, then astronauts, becoming one of America's most dependable launchers.

other. The difference between them was that almost everything the United States did was public, including the failures; the U.S.S.R. took care to announce only its successes.

In April 1958, America's first Thor-Able blew up 10 seconds into its flight. In May—in time, Khrushchev hoped, to persuade Italians to vote Communist in their upcoming elections—Korolev launched Sputnik 3, 1.3 tons of it, almost warhead size. This was a strategic coup, but made no measurable difference to the Italian election results.

The following month, Roy Johnson told rocket-builders Convair: "We've got to put something big up." "Well," came the reply, "we could put the whole Atlas in orbit." Johnson agreed, and imposed tight secrecy—only 88 people knew the true purpose of the upgraded Atlas tests.

On July 29, Eisenhower moved to cut through the burgeoning bureaucratic tangle by ordering the creation of a new federal agency: NASA, the National Aeronautics and Space Administration. At its core was the small but long-established and well-respected National Advisory Committee for Aeronautics, which had designed the now standard-shaped nose cone in its laboratory in Langley, Virginia. When NASA came into operation on October 1, 1958, it grew fast, taking over the Vanguard project and the Jet Propulsion Laboratory. It also started work on the new Goddard Space Flight Center, near Baltimore, for science-based, unmanned spacecraft.

A voice from space

NASA also advanced the growing work on placing a man in space—an idea first seriously proposed by von Braun in what he called Project Adam, which suggested putting a man on top of a Redstone and shooting him 150 miles up. No one took him seriously, partly because the Air Force spearheaded high-altitude flight and people assumed that the first spacemen would be pilots flying rocket planes. Even when NASA took over the man-in-space work—Project Mercury—no one gave much thought to why a man should be in space. It was enough that the Soviets might do it first.

In August 1958, Atlas made a 3,000-mile flight. The following month, a try for lunar orbit blew up, coinciding with the explosion of a Soviet R-7 aimed at the Moon.

The next lunar "launch-window" would come a month later, and both sides intended to take advantage of it. By then, refinements had been made to Thor-Able. When it blasted off for the Moon in mid-October, all seemed well until the third stage—containing the Pioneer 1 spacecraft designed to orbit the Moon—failed to separate cleanly and veered off at an angle.

Korolev knew of the American launch; he also knew that his more powerful rocket could reach the Moon first, if all went well. It didn't: the R-7 exploded after only 104 seconds, scattering fragments over the frozen steppes of Kazakhstan. A salvage team, living in tents and eating whatever they could shoot, tracked it down to find out what had happened. It took two months to identify the problem—variations in pressure in the liquid-oxygen lines—and get it right.

The Thor-Able had fared better; even off-course, it might have escaped Earth orbit. To do so, it needed to reach 24,600 mph, and it had failed by just 500 mph, a mere 2 percent of the required velocity. Even so, it had reached a record distance of 70,000 miles from Earth. In December, the second of von Braun's Juno rockets, carrying Pioneer 3, cut off a few seconds too soon and reached almost the same distance.

Three weeks later, on December 18, came an American success, at last: an Atlas launch into orbit. The next morning, as people switched on their radios across the nation, they heard Eisenhower delivering a Christmas message: "This is the President of the United States. Through the marvels of scientific advance my voice is coming to you

AIMING FAR Von Braun (center) looks tense as he awaits the launch of Pioneer 4 at Cape Canaveral in 1959. Pioneer 4 became the first American craft to escape the Earth's gravity.

from a satellite circling in outer space. My message is a simple one. Through this unique means I convey to you and to all mankind America's wish for peace on earth and good will toward men everywhere." One of Eisenhower's predecessors, Theodore Roosevelt, had a foreign-policy maxim: "Speak softly, and carry a big stick." Eisenhower's simple good wishes had a very big stick behind them. Unequivocally, America was back in the game.

More big sticks were being built: nuclear submarines with solid-fuel Polaris missiles, U-2 spy planes, and a new series of spy satellites code-named Discoverer. Unlike other spy satellites, which took an east-west course around the Equator, the new Discoverer satellite was to fly a north-south polar orbit. But a launch from Florida would take the rocket over populated areas. It was therefore re-sited at the new Vandenberg Air Force Base in California. The first stage would be a Thor, the second stage an Agena, from a sister project to that of Thor-Able.

Beyond the Moon

In January 1959, an R-7 lifted off carrying a new 800-pound lunar spacecraft, Luna 1, which was intended to crash into the Moon. Everything worked perfectly, and the probe became the first man-made object to escape Earth's gravity. However, it then missed its target and traveled on to join the planets in its own orbit around the Sun—a milestone in itself. Korolev's bosses were ecstatic, and both were eager for more successes.

Two months later, the United States matched the Soviets when a Juno rocket catapulted Pioneer 4 past the Moon into solar orbit, and tracked it to a distance of 400,000 miles.

NASA was growing fast. It had a range of rockets that could launch anything from small satellites to a lunar mission, once development was more advanced. But why, after all, put a man in space? Why go to the Moon? Why the need for Saturn?

There were no strong rational arguments, but it was simply inconceivable to abandon the idea when the Soviets were forging ahead so fast. Eisenhower made a decision: NASA would take over Saturn and von Braun's entire rocket group. Von Braun became a civilian for the first time in his life since joining the German defense force in 1932. The Army's Redstone Arsenal near

THE ORIGINAL SEVEN

When they were chosen in 1959, the first seven trainee American astronauts were unknown. Soon they were stars, with publishing contracts, promotional deals, and in demand for public events everywhere. All were dedicated professionals making a unique contribution to manned space flight.

Air Force Captain L. Gordon Cooper
Born in 1927. Flew in Mercury 9 and Gemini 5. He developed a reputation as the most laid-back of the astronauts, especially after he fell asleep atop a fully fueled Atlas booster.

Air Force Captain Virgil "Gus" Grissom
Born in 1926. Flew on Mercury 4 and Gemini 3. Small, powerful and perhaps the most fiercely competitive of this competitive group. He died during training in an Apollo spacecraft fire in 1967.

Air Force Captain Donald K. "Deke" Slayton
Born in 1924. Perhaps the most gifted pilot of the Seven, but was grounded with a heart condition. He then served as Apollo's Director of Flight Crew Operations. He flew on Apollo-Soyuz in 1975, and died of a brain tumor in 1993.

Navy Lieutenant M. Scott Carpenter
Born in 1925. Flew on Mercury 7. He reputedly wasted fuel on his mission, crucially mistimed his re-entry, landed 250 miles off-target, and was never again selected to fly.

Navy Lieutenant Commander Walter M. Schirra
Born in 1923. Flew on Mercury 8, Gemini 6 and also Apollo 7. Known for his good-natured bluster, he became notorious for his bad temper—the result of a cold—on the Apollo 7 flight.

Navy Lieutenant Commander Alan B. Shepard
Born in 1923. Flew in Mercury 3 and Apollo 14, thus becoming the first American in space and the only member of the Original Seven to reach the Moon. A genius with machinery, his confidence inspired awe.

Marine Lieutenant John H. Glenn
Born in 1921. He flew on Mercury 6, becoming the first American in orbit in 1962. He then proved so adept at public life that he retired from the Marine Corps in 1965 and went into politics. In 1998 he went into space again at the age of 77.

SUITED AND READY America's Original Seven were (back row, left to right) Alan Shepard, Gus Grissom and Gordon Cooper; (front row, left to right) Walter Schirra, Deke Slayton, John Glenn and Scott Carpenter.

THE FAR SIDE OF THE MOON

Soon after the Moon was formed, tidal forces exerted by the Earth's oceans slowed its spin until it presented just one face to the Earth. As a result, the two sides developed differently. The side facing the Earth is scarred with vast dark areas that were once thought to be seas, and are still known as maria, or "seas." They were probably caused by huge rocks falling onto the Moon's surface after separation from the Earth. The far side, by contrast, has far fewer maria (though one, Mare Orientale, has a ring of mountains that suggest a near-vertical impact).

There are craters everywhere, peppering the surface as evenly as they do on the near side. The far side, however, has one unique feature, obvious in the first pictures taken by Luna 3 in 1959—an impact crater so large that it can almost be classed as a "sea." Named Tsiolkovsky, after the Russian rocket pioneer, it has a dark lava floor and a high central peak typical of many impact-craters. At 93 miles across, it is one-and-a-half times as large as the near side's biggest crater, Copernicus.

HIDDEN FACE REVEALED   The first photographs of the Moon's far side, taken by Luna 3, were blurry and indistinct. Nevertheless, they caused huge excitement.

Huntsville became the civilian Marshall Space Flight Center. With the Navy's Vanguard rocket already inside NASA, the decision made inter-service rivalry a thing of the past.

In April 1959, the Armed Forces were asked for the names of men who could match the demanding requirements of a new project, called Mercury, to place a man in Earth orbit. They could not be taller than 5 feet, 11 inches, and they would, inevitably, come from an elite group: the very best of the nation's jet pilots. Of 500 on the short list, seven were chosen, all of whom would become household names. Six would fly in Project Mercury, while one, Alan Shepard, would be the first American in space and one of the men who would walk on the surface of the Moon. They were all treated, from the start, like royalty, and were united by a unique bond.

The dark side of the Moon

In September 1959, Khrushchev was due to visit the United States, and was eager for another coup. Korolev obliged by launching Luna 2 directly at the Moon, and to prove the Soviet Union's peaceful intentions, he arranged for the journey to be monitored by the Jodrell Bank Radio Telescope. The telescope's founder, Sir Bernard Lovell, was able to follow the probe's transmissions and pinpoint the exact moment when the probe crashed in the Mare Imbrium on September 13.

Korolev's next success, on the second anniversary of Sputnik 1, was to fire Luna 3 on an even more accurate trajectory that would allow it to enter the Moon's gravity without being trapped. Accelerated by the Moon's gravity, Luna 3 whipped around the back of the Moon,

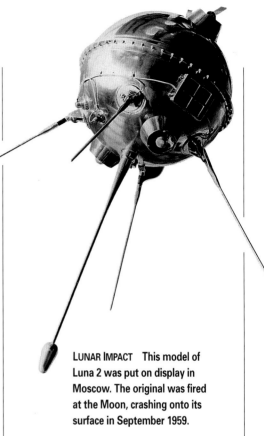

LUNAR IMPACT   This model of Luna 2 was put on display in Moscow. The original was fired at the Moon, crashing onto its surface in September 1959.

raced out of lunar gravity, and back toward the Earth. More than that, it took 29 photographs of the dark side of the Moon, which were processed on board. The prints were scanned by a fax at 1,000 lines per frame. But the probe lacked the power to transmit the images back to Earth from the Moon itself, so Korolev and his team would have to wait until the craft was close to home, a week after its launch.

No one knew how close it would have to be. At the receiving station in the Crimea, where local roads were closed to traffic, and ships were told to keep radio silence, Korolev's team sent signal after signal to the probe. Four tries, and nothing came through on the monitor. The fifth time, a white disk showed up. "At least we know the far side is round," someone joked. Then, as the signal strengthened, more detail appeared. In the end, 17 images emerged, fuzzy perhaps, but a genuine scientific triumph. For the first time, people on Earth saw the Moon's hidden face. They named its grandest features in honor of their own scientists—Tsiolkovsky, Lobachevsky, Lomonosov—together with Maxwell, Edison, Verne and Pasteur from other nations.

In April 1960, the research that would take men into space picked up momentum. Von Braun tested a prototype Saturn, its eight engines together producing a thrust of

DISASTER ON THE STEPPE

In mid-October, 1960, Khrushchev was visiting the United Nations in New York. Soviet tracking vessels were observed leaving their ports, and another space coup of some kind was expected. But nothing happened. Then, on October 25, it was announced that Field Marshal Mitrofan Nedelin, missile chief in the Soviet military, had died "in a plane crash." Over the years, reports leaked out linking these two enigmatic events. It was said that a catastrophic event had occurred at the Tyuratam launch site.

There was no hard information, and only a few rumors. One account of what happened appeared in 1976 in the British weekly *New Scientist*, written by the Soviet biochemist Zhores Medvedeev, which revealed that Nedelin and dozens of others had been killed while attempting to repair an ignition failure during a countdown. Clearly, this had been the worst disaster in the history of rocketry.

Further details emerged only when the Soviet empire collapsed in 1989. The world then learned that Nedelin had been masterminding a new ICBM, the R-16, a replacement for the R-7. Developed under Korolev's protégé, Mikhail Yangel, it was planned to carry a 10-megaton bomb 6,500 miles. Unlike the R-7, it could be hidden away in silos, and it used fuels that could be stored, rather than having to be manufactured on the spot just before a launch. One of the new fuels was nitric acid. The first trial flight, at Yangel's own center at Tyuratam, was scheduled to coincide with Khrushchev's New York visit.

During the countdown, problems developed. Instead of draining the fuel, which would mean postponement, Nedelin ordered repairs immediately. An error caused the second stage to ignite, and the flames pierced the first stage, which turned into a fireball, releasing nitric acid in vapor and flames.

"People were running toward a special overhang," wrote a survivor. "But in their path was a strip of freshly laid asphalt. It immediately burst into flames, trapping many in the hot, sticky mess. Later, outlines of men and non-combustible items like metal coins, bunches of keys, pins and belt buckles could still be seen at this site. The most terrible fate fell upon those who were in the upper levels of the servicing gantry. People were enveloped by the fire and burst into flames like intense flares. Later, scorched corpses hung everywhere on the barbed wire that surrounded the site."

Among the dead, numbering over 100, was Nedelin himself. Yangel was having a cigarette in a protected area, and survived. Nedelin was buried in the Kremlin Wall, with full military honors and no further explanation. Only the survivors knew the truth.

---

580 tons, and creating a roar that could be heard 100 miles away. Rocketdyne started to build new facilities at Edwards Air Force Base in the Mojave Desert in California.

Still, no one could say exactly why it was necessary. Eisenhower ordered a study on the aims, costs and benefits of the project. The aim, of course, would be a manned journey beyond the Earth, around the Moon. Already NASA was eating up $1 billion a year. The cost of the new venture would be $8 billion, and that didn't even include a lunar landing. When Eisenhower and his colleagues were told this, they were bewildered. Somebody remarked: "This won't satisfy everybody. When they finish this, they'll want to go to the planets." Everyone laughed at the foolishness of the idea.

But Saturn was not canceled, as the perceived threat from the other side intensified. Korolev was firing off one R-7 a week, and the Soviets were known to be training their own astronauts. Intelligence also reported that the Soviets would have 450 ICBMs by

1963, twice as many as the United States. There was, in the catch phrase of the day, a "missile gap," which needed to be closed.

It got worse. On May Day, 1960, a Soviet surface-to-air missile brought down an American U-2 spy plane, which was not only a propaganda coup for Khrushchev, but also proof that the U-2 was vulnerable. The United States needed cameras in orbit.

Success with Discoverer

In August 1960, after a dozen failures over the previous year and a half, a Discoverer spy satellite finally worked perfectly. It re-entered as planned and floated down by parachute into the Pacific. A few days later, Discoverer 14 went into orbit with a working camera on board. After eight orbits, it had

ONE MAN AND HIS DOG Standing at the entrance to a pressure chamber are a Soviet astronaut and a dog wearing its own specially modified space suit. Dogs played an important role in Soviet research into the likely effects of space flight.

photographed one-fifth of the Soviet Union. It then ejected its film capsule, which was caught on a parachute in midair by an Air Force plane. When processed, the pictures formed eight broad panels with resolution down to 30 feet.

This and other Discoverer missions revealed startling information. There was no massive superiority of Soviet ICBMs. The only threat the Soviet Union could mount was by means of Korolev's R-7, which might have been a good space launcher but was far too cumbersome and vulnerable to be used as an ICBM. The Soviets were developing a replacement R-16, which could be hidden away in silos, but was still a long way from being ready.

In brief, the United States was not behind at all in the development of ICBMs, and was way ahead in other areas. But only a few

WEATHER EYE The first weather satellite, TIROS 1 (left), was launched on a Thor-Able rocket in April 1960 (below). Using two cameras, it took some 23,000 photographs of weather systems around the globe before failing two and a half months later.

insiders knew the truth, and they kept it under wraps because no one wanted a hint of complacency in the face of continuing Soviet saber-rattling, jibes, boasts and propaganda coups. The last included satellite launches in the summer of 1960, which placed two dogs in orbit and returned them safely to Earth.

Satellite communications

Though long delayed, proof of American dominance would come sooner or later. It was already there for those who could see it.

In April 1960, NASA sent up TIROS 1 (Television and Infra-Red Observation Satellite). For the next two and a half months it sent back 23,000 black-and-white images, tracking storms and proving itself a vast improvement over observation aircraft, which were expensive and easy victims of the bad weather they were supposed to track.

Another series of satellites was used for communication. The idea initially derived from the needs of nuclear submarines, which needed to know their positions without surfacing and without using sonar, which might give them away. The solution was a radio satellite in a known orbit. All the submarine had to do was listen and interpret in order to establish its position.

Finally, in a genuine step into the unknown, Pioneer 5 headed away from Earth in March 1960 to test an array of solar cells. Its cells generated only 5 watts, enough to light a single Christmas tree bulb, but they proved reliable. Pioneer's radio signals were audible for three months, until the little probe vanished into deepest space 22 million miles from home.

Up to 1960, almost all space research related to unmanned flight. No one could see a purpose in manned space flight. Certainly, Eisenhower saw no reason to spend vast sums on a Moon mission. That essential sense of purpose would come only from a restatement of policy, which would once again match dreams and reality.

DAWN OF A NEW ERA

AFTER THE SOVIET UNION SUCCEEDED IN PUTTING THE FIRST MAN INTO SPACE, THE RACE FOR THE MOON BEGAN IN EARNEST

In January 1961, John F. Kennedy took over from Eisenhower as President, and proclaimed that "the torch [had] been passed to a new generation." Though he had previously not known much about space research—there were no big rocket contractors in his home state of Massachusetts—Kennedy used the issue to great effect during his 1960 presidential campaign. "The first canine passengers in space who safely returned were named Strelka and Belka," he warned wryly, "not Rover or Fido." By the

WORLD HERO Yuri Gagarin lies strapped to his seat in the Vostok 1 space capsule (below). He was given star status not just in the U.S.S.R. but throughout the world. His face appeared on everything from stamps (inset) to billboards.

time Kennedy came to office, the Soviets had recovered another two dogs from orbit. A manned launch could not be far behind, and Kennedy knew that he would have to deal with the consequences. Indeed, for some weeks, insiders in Moscow had heard that a manned flight was imminent.

The first man in space

In January 1960, a new cosmonaut training center called Zvezdny Gorodok ("Star City") had been set up near Moscow. Here, six young pilots were subjected to the imagined extremes of space flight, such as massive G-forces and weightlessness—the former simulated by a huge centrifuge, and the latter by a jet diving in parabolic flight. One trainee, a 27-year-old lieutenant named Yuri

THE FIRST MAN IN SPACE

After his historic flight, Yuri Gagarin provided the first description of Earth as seen from space:

"I could see the horizon, the stars, the sky. The sky was completely black, black. The magnitude of the stars and their brightness were a little clearer against that black background. I saw a very pretty horizon, and the curvature of the Earth. The horizon is a pretty, light blue. At the very surface of the Earth, a delicate light blue gradually darkens and changes into a violet hue that steadily changes to black.

"In my flight over the sea, its surface appeared grey, and not light blue. The surface was uneven, like sand dunes in photographs. I ate and drank normally. I could eat and drink; I noticed no physiological difficulties. The feeling of weightlessness was somewhat unfamiliar compared with Earth conditions. Here, you feel as if you were hanging in a horizontal position in straps ...

"When I emerged from the shadow of the Earth, the horizon looked different. There was a bright orange strip across it, which again passed into a blue hue and once again into a dense black color ...

"[On re-entry] I was an entire corps de ballet. Head then feet rotated rapidly. Everything was spinning around. First I saw Africa, then the horizon, then the sky. I only barely managed to hide my eyes from the Sun [by drawing his shade] ...

"A bright crimson light appeared along the edges of the shade. I felt the oscillations of the craft and the burning of the heat shield. It was audibly crackling; either the structure was crackling, or the heat shield was expanding as it heated. I felt that the temperature was high. Then the G-load began to steadily increase. It felt as if it were 10 Gs. There was a moment for about two or three seconds when the instruments began to become fuzzy. Everything seemed to go gray. I strained to see, and that helped ..."

[The hatch blew off, his ejection seat fired, and he found himself descending by parachute over the Volga. He landed in a field] "well plowed, very soft, and it hadn't even dried up yet. I did not even feel the landing. I didn't even realize I was already standing on my feet. I saw that everything was intact. That meant I was alive and well."

Gagarin, showed greater endurance and equanimity than the others, and accordingly spent more and more time in a mock-up of Korolev's Vostok ("East") space capsule.

April 12, 1961, was a fine spring morning on the Kazakh steppes. Gagarin and his back-up, Gherman Titov, had spent the night in the wooden cottage that Korolev used near the launch pad. After a breakfast of pâté, bread, black currant jam and coffee, they each dressed in woolen long johns, a pressure suit and bright orange overalls that would make them easy to see after landing. Finally came the communication hat and the pressure-resistant helmet.

They were driven to Korolev's R-7 rocket, on top of which sat the Vostok spacecraft, which had proved its reliability with dogs, rats, mice and insects. If all went well, Gagarin would have nothing at all to do, except comment. Korolev rode with him to the top of the rocket and helped him in.

Almost an hour later, at 9:07 a.m., Gagarin was off, without fuss. The noise, he reported, was no worse than that of an aircraft; the acceleration was bearable, though "it was somewhat difficult to talk, since all the muscles of my face were drawn." After two minutes, the nose cone, or fairing, fell away and Gagarin could see the view. "I

could see forests, rivers, ravines. I couldn't tell exactly where it was. I think it was the Ob or the Irtysh," he reported later. The rocket, still under power, continued to carry him up into orbit around the Earth.

Gagarin radioed back brief descriptions of his experiences, and even made a few notes until he let go of his pencil and saw it float away out of reach. In case his retrorockets failed, he had food for ten days—long

SAFELY HOME The capsule of Vostok 1, in which Gagarin flew once around the Earth, lies in a farmer's field near the Volga River, with its parachute close beside it.

enough for his low orbit to decay and allow him to fall back to Earth. His temperature, blood pressure, heartbeat and breathing all remained normal. *Homo sapiens* could apparently survive in space unharmed.

After 68 minutes the retrorockets fired, beginning the process that would drop the capsule back into the atmosphere. There was a moment of tension when Gagarin's descent capsule refused to separate from the instrument section, but the heat of re-entry forced them apart.

At 23,000 feet he was ejected. His parachute opened and he floated down into a newly plowed field not far from the Volga River. He was seen by a peasant woman, Anna Takhtarova, and her six-year-old granddaughter. After examining his scorched capsule, Gagarin explained where he had come from. Anna refused to believe him until some tractor drivers appeared who knew about the launch. Then a truckload of soldiers arrived to pick Gagarin up and salute him as a newly promoted major.

YURI GAGARIN: SOVIET COSMONAUT AND HERO

Yuri Gagarin was born in 1934 on a collective farm west of Moscow. He went to a technical school in Saratov, on the Volga 150 miles north of Stalingrad. He learned to fly, joined the air force and graduated from officer training in 1957. He married and lived with his wife and two baby daughters in a small apartment outside Moscow, where he and 20 others were training as cosmonauts.

After his momentous space flight, Gagarin became a national and international hero with ceremonial duties that were arduous but uncreative. After two years of fulfilling these duties, he demanded more action and was reassigned as a cosmonaut. Gagarin probably would have flown in space again, but he was killed in March 1968 when his training jet crashed into a forest. He was given a state funeral and his ashes were buried, as befitted a national hero, in the Kremlin Wall.

INTERNATIONAL STAR Two weeks after his space flight, Gagarin is given a hero's welcome in Prague during a triumphant, but hectic, tour of Eastern Bloc capitals.

Kennedy's "New Frontier"

This new Soviet success, accompanied by more banner headlines, more breast-beating on Capitol Hill and yet another massive celebration in Red Square, was the last straw for Kennedy. "Now let's look at this,"

January 20 John F. Kennedy inaugurated as President

April 12 Yuri Gagarin becomes the first man in orbit

May 5 Alan Shepard becomes first U.S. astronaut in space

July 21 Gus Grissom in Mercury 4 is the second U.S. astronaut in space

August 6 Gherman Titov (U.S.S.R.) orbits the Earth 17 times

ESCAPING EARTH'S GRAVITY

he said to his glum advisers two days later. "Is there any place we can catch them? What can we do? Can we go around the Moon before them? Can we put a man on the Moon before them?"

There was only one hope: not to compete for today and tomorrow, but to look beyond that, to something bigger, something new, in which America had a chance of success. The prospect of a program costing tens of millions of dollars was awesome, but Kennedy had made a decision. He thanked the men for coming, then spoke briefly to his closest associate, Ted Sorensen. Five minutes later, Sorensen came out of the meeting. "We're going to the Moon," he said.

It was a gut decision, but one that could withstand hard political and economic analysis. Government involvement in massive projects had revitalized the American economy in the past. Lincoln had backed the first transcontinental railroad. The Panama Canal; the great dams of the Depression years; the postwar interstate highways; the bombers that were the basis of the new big jets; atomic power (at a time when it seemed a clean alternative to coal)—all were rooted in federal initiatives. They had created work and welded the country together. Moreover, Kennedy was the first Democrat to succeed a Republican since Roosevelt defeated Hoover in 1933. Roosevelt had spent his way out of the Great Depression with grand projects. Kennedy could use something equally grand to lift the country from another sort of depression.

Such a goal suited his talents. A revival of the American spirit demanded idealism of a high order, and Kennedy, the very image of youthful energy, could inspire like no other. Like other ambitious leaders, he had his eye on history, and history had given him a supreme moment in which to define a new direction. He seized the moment and pointed the way with spellbinding oratory.

Space, Kennedy said, was the "New Frontier"—a phrase designed to appeal to something deep in the American psyche that had been defined by the opening of the western frontier. But this new frontier was infinite, in dimension and in theme. Kennedy could look beyond mere Cold War politics, beyond even the immediate needs of the American nation, to something

Having grasped the main principles of rocketry—that a rocket is projected upward by reaction to the thrust of exhaust gases produced by burning fuel, and that as the fuel burns, the rocket gets lighter and acceleration increases—scientists then had to match them with the laws that govern any object within the Earth's gravitational reach. These laws were first described by Isaac Newton in the 17th century. It was Newton who realized that the principle of gravity is universal, causing apples to fall to the ground, holding the Moon in orbit around the Earth, joining the Earth and other planets to the Sun, and binding the stars into the great swirl of the Milky Way.

The strength of the gravitational pull exerted by any planetary body, including the Earth, decreases with distance by the square of an object's distance from it. The strength of a planet's gravitational field is also dependent on its mass, which is why the Earth has much greater gravitational "pull" than the Moon, and why objects on Earth weigh more than they do on the Moon. The Earth's gravitational field creates an "aura" reaching some 200,000 miles into space, though its effects are felt even farther away than that.

For rocketeers, the nature of the Earth's gravitational field creates an initial problem, which then turns to their advantage. The problem is that because the force of gravity is strong near the Earth's surface, huge lifting power, or thrust, is needed to get a rocket off the ground. The advantage is that once a rocket attains enough speed, no further acceleration is necessary and the rocket engines can stop burning. This happens because the strength of gravity decreases at an ever-increasing rate with distance from the Earth's surface.

The speed, or "escape velocity," required to take a spacecraft beyond Earth's gravitational field is the same as that of a rock falling toward Earth from the fringes of the Earth's gravitational reach: 6.8 miles per second, or 24,600 mph. At this speed, without its engines, a rocket heading away from Earth will slow, like a coasting car reaching the top of a hill. Then, even if it is only moving at walking pace, the rocket will gradually "fall," as if down another slope, into the gravitational field of the Sun or Moon. If it is traveling at a speed slower than escape velocity, the rocket will slow, come to halt, then fall back to Earth with ever-increasing speed.

If, instead of rising or falling vertically, an object, such as a satellite, moves sideways within the Earth's gravitational field, the object's path forms a curve. If this creates an outward pull (as in a centrifuge) that matches the downward pull of Earth's gravity, the object will "fall" in a complete circle, and continue to orbit Earth in the vacuum of space forever. The principle is the same around any planet or moon.

The Moon has only a fraction of the mass of the Earth, so objects can orbit the Moon at 3,800 mph and escape its gravitational pull at 5,370 mph.

At an altitude of 22,500 miles above the surface of the Earth, this communications satellite has a velocity of 6,935 mph and orbits the Earth once every 24 hours. At low altitude, objects need a velocity of 18,000 mph to stay in orbit.

A rocket must reach a speed of 24,600 mph to escape the Earth's gravitational attraction completely—for example, to go to the Moon.

GETTING AWAY Rockets are launched near the Equator, in the same direction as the Earth's spin—to the east—to help them gain as much speed as possible. Once they pass a critical point, they are drawn toward the Moon. Returning to the Earth, a spacecraft enters orbit in the same direction as Earth's spin, before re-entering the atmosphere for landing.

By taking off near the Equator, rockets use the speed of the spinning Earth to help "throw" them into space.

greater still: humanity's need for knowledge, for growth, for grandeur.

Having made his decision, Kennedy was confronted with a sudden crisis requiring strong leadership. At one level, the Cold War existed right on his doorstep, since Cuba had fallen to Fidel Castro's Communists two years previously. When he took office, Kennedy inherited a wild CIA-backed scheme to destabilize Castro by sending in an invasion force of exiled Cubans. When the brigade landed in the Bay of Pigs on April 17, only a week after Gagarin's flight, Castro's troops were ready. The invasion was a fiasco. The exiles were killed or captured, and in effect held for ransom. A year later, 1,100 were set free in return for $53 million. Kennedy found himself severely criticized by opponents at home and abroad.

Alan Shepard's flight

On April 19, Kennedy called his vice president, Lyndon B. Johnson, to his office and asked him if there was any space program "which promises dramatic results which we could win." It was a good time to ask such a question, as the world would soon know.

On May 5, as part of Project Mercury, Alan Shepard was strapped into a Redstone rocket for a suborbital lob. It took an unforeseen length of time—four hours—to get through the pre-flight checks, by which time Shepard needed the sort of relief for which no provision had been made: he was forced to relieve himself inside his space suit. He then did his best to reassure his nervous

FIRST AMERICAN IN SPACE Alan Shepard is winched from the sea (left) after his brief trip into space. As his mission badge (inset) shows, his capsule bore the patriotic name *Freedom 7*, which also referred to the Original Seven team of astronauts. Three days later President Kennedy honored Shepard by presenting him with a medal at the White House (below).

ground controllers: "I'm cooler than you are. Why don't you fix your little problem and light this candle?"

It was only a 15-minute flight, carrying Shepard up 115 miles, but it made him the first American in space. Unlike the Soviet launches, this was a very public event. In Indianapolis, a judge halted a trial so that everyone could watch the event unfolding on television—the television in question being the one stolen by the defendant. Drivers pulled off the road to hear the commentary. They heard Shepard's voice transmitted live with a strange, tinny sound: "What a beautiful view. Cloud cover over Florida . . . I just saw the Andros Islands, identified the reefs."

It was just what the nation needed. After a ticker-tape parade in New York, Shepard was invited to the White House to receive the nation's congratulations. It was just what Kennedy needed, too, for it gave proof of America's technical prowess to do what he had in mind: to set an achievable goal, but one far enough ahead that the Soviet's existing lead gave them no long-term advantage.

Kennedy's vision

On May 25, 1961, as NASA prepared for the launch of Mercury 4, Kennedy delivered what was in effect a state of the union message to a joint session of Congress. He intended to build up the armed forces; he urged the building of nuclear fallout shelters; he would increase foreign aid. Then, toward the end of his speech, he made the crucial link between space achievements and "the minds of men everywhere." At last came the ringing words that wrote America's new agenda for space:

"Now it is time to take longer strides— time for a great new American enterprise— time for this nation to take a clearly leading role in space achievement, which . . . may hold the key to our future on Earth.

"I believe that this nation should commit itself to achieving the goal, before this decade is out, of landing a man on the Moon and returning him safely to Earth. No single space project in this period will be more exciting, or more impressive to mankind, or more important for the long-range explo-

ration of space; and none will be so difficult or expensive to accomplish."

Kennedy's timing was impeccable. The United States might be behind in space, but it was ahead on the ground, with more than 100 ICBMs (to the Soviet Union's 20) and some 1,300 long-range bombers (to the

AMERICA'S SECOND SPACEMAN Gus Grissom suits up in his capsule (inset, middle) for the flight of Mercury 4. Grissom's badge (inset, top) shows that the capsule was named *Liberty Bell 7*. Liftoff in the Mercury-Redstone rocket (main picture) was delayed for two days, and the flight almost ended in disaster.

RULES OF SPACE: RE-ENTRY

In the early days of rocketry, scientists assumed that rockets re-entering the atmosphere from space needed sharp noses to cut through the air in order to avoid overheating and burning up. However, research in the early 1950s showed that exactly the opposite was true. The missile or spacecraft needed to absorb and re-radiate the heat generated by the friction of re-entry, and the best design for this purpose was a blunt nose, the outer layers of which were allowed to burn away like a meteor. This approach was also adopted for re-entry capsules in the early days of manned flight.

To protect a manned capsule, it had to be designed with two issues in mind. First, the heat shield had to be strong enough to withstand the tremendous heat of re-entry; second, the spacecraft had to be aerodynamically stable so that it would remain balanced against the tide of air rushing past it, like a marble on a fountain. It then had to be correctly positioned in space before entering the atmosphere, so that it led with its angled heat shield. Failure to do so would turn the spacecraft into a fireball.

The angle of re-entry is crucial. If it is too steep, the capsule will burn up completely; too shallow, and the spacecraft will skip off the upper atmosphere like a stone bouncing on water (although with Apollo, this effect was deliberately used to slow the capsule before final re-entry).

The heat created by the friction of re-entry presented a final problem, for which there was no solution. Scientists found that atoms of air flowing around the capsule were ionized by the heat, temporarily cutting radio contact. For about four minutes as the capsule slowed, there would be a radio blackout, imposing extra tension on those waiting anxiously below for a sign that their crew had come through safely.

The capsule design also had to take account of the need to slow the craft further in the lower atmosphere, and this effect was achieved through the release of parachutes. To land safely, the capsule would also have to be shielded from impact on the ground by cushions; or if it landed in the sea, it would have to act temporarily as a boat, bobbing on the waves until ships and helicopters could come to the rescue.

CRITICAL TIMING This diagram shows one of the major difficulties of getting home safely from space. Timing is crucial: if the spacecraft enters the atmosphere too steeply, it will create too much heat and burn up (left); too obliquely, and it will bounce away (right); the angle needed for success (center) is critical.

and western sectors, and divided families. There was virtually nothing that the West could do, and Khrushchev drew strength from the success of his gamble.

Next into space, in July, was Gus Grissom, a veteran of 100 combat missions in Korea. Until the landing, his flight mirrored Shepard's. Then everything went wrong. When Grissom splashed down in the Atlantic, his capsule's hatch blew out too early. The rescue helicopter couldn't lift it and had to cut it loose, while Grissom himself, struggling in his waterlogged space suit, almost drowned before he was rescued. No blame was attached to him officially—the blowout was written off as a freak accident—but Grissom felt the burden of his "failure," until his successful two-man flight in 1965.

Two weeks later, the Soviet Union came back with yet another record. Gagarin's back-up, Gherman Titov, flew for a whole day in space, making 17 orbits of the Earth. Titov suffered from nausea and picked up an ear problem that kept him grounded thereafter. But it was another great step forward, and Khrushchev's self-confidence hardened into a brittle and unrealistic attitude that would lead to near-catastrophe within a year.

Soviet's 58). In addition, the nation had rockets positioned in Britain, Italy and Turkey.

The facts were one thing, but the emotional impact of Kennedy's statement on his Soviet counterpart was quite another. In June, Kennedy and Khrushchev met in Vienna. Kennedy put on a friendly face. Khrushchev, a man of bombast as well as peasant cunning, interpreted this as weakness and formed the impression that he

could outface Kennedy in a crisis. That summer, he gave orders for the building of the Berlin Wall, which cut the city in two, ended travel between the city's eastern

ENDURANCE RECORD Gherman Titov, seen here in training gear, set a record of 17 orbits in August 1961—and became the first person to suffer from space sickness.

PREPARING FOR THE BIG ONE

TO GET A MAN TO THE MOON, BOTH RUSSIA AND AMERICA HAD TO DEVELOP TECHNIQUES FOR MANEUVERING AND DOCKING IN ORBIT

President Kennedy's promise to open a new frontier in space became a reality with remarkable speed. NASA, backed by $2.5 billion, had begun a giant sweep of industrial and scientific talent, transforming whole industries and taking on an army of workers and contractors.

Major new facilities had to be planned—but where? California was a natural choice. But two key members of the Kennedy administration—Vice President Lyndon Johnson and Albert Thomas of the House Appropriations Subcommittee, who had supervision of NASA's budget—were from Texas. They wanted to boost their local economy and argued that Texas, like California, is ice-free, but also has more water, allowing year-round transportation by canal and river for the immense pieces of equipment. The South would become NASA's base, with Houston featuring prominently.

NASA's plans for building a Moon rocket accelerated rapidly. The agency announced an extension to the Cape Canaveral launch

FLIGHT TRAINING Among the many training devices developed by NASA was this simulator, which taught astronauts how to pilot their craft in space.

site in Florida, which involved taking over Merritt Island, to the west of the Cape. There NASA would build the world's largest enclosed space, the Vehicle Assembly Building, in which four Moon rockets could be assembled side by side. A huge ship-assembly building near New Orleans, with 43 acres of floor space, would provide room for the assembly of the rocket's first stages. On the Pearl River, 35 miles away, von Braun found an ideal test site for rocket motors that would deafen the locals. There, above the swamps, cypress groves and swarms of mosquitoes, would spring up a 400-foot high test stand, itself founded on 1,600 pilings 100 feet high.

Finally, Project Mercury's Space Task Group, which would grow into Mission Control for the Moon venture, needed a new base. A company called Humble Oil, happy to please Southern Democrats, offered 1,000 acres of cow pasture south of Houston. This would become the site of the Manned Spacecraft Center. Over the next few years, all these facilities would be used to build experience in one and two-man missions that would eventually

give way to the three-man voyage to the Moon, code-named Apollo. But first a number of issues needed to be addressed.

The first step was Earth orbit, in which all the major systems and techniques would be tested. Space suits needed to be developed, not just for walking on the Moon, but for making repairs and handling instruments in flight. Maneuvering in order to rendezvous, docking and space walking all demanded the development of new skills.

Of the several rockets available as launch vehicles for the early one-man flights, only the Atlas ICBM had the power to put a

TRAINING CAPSULE Strapped securely into a working mock-up of the Mercury capsule, John Glenn goes through the procedures he would perform as the first American in orbit.

Mercury spacecraft into orbit. But the Atlas was not yet reliable. In three unmanned launches, two had exploded. Operations Director Walt Williams favored the Titan, a view that galvanized Atlas's manufacturers, Convair, into a frenzy of revisions. In September 1961, an Atlas rocket put an unmanned Mercury into a single orbit in a perfect operation. In November, a monkey named Enos took a flight. Again, everything worked to perfection.

John Glenn goes into orbit

The original plan was that after Shepard and Grissom, the other five of the Original Seven would all have suborbital lobs. Gagarin's flight short-circuited that idea. Instead, the next in line, John Glenn, would go straight into orbit.

Disdaining the tough, hard-drinking image cultivated by other test pilots, Glenn

STARSHIP TROOPER John Glenn, hero of the Mercury 6 mission, looks every bit the resolute explorer. As his badge (inset) shows, he named his capsule *Friendship 7*, seen below lifting off on top of a mighty Atlas rocket.

was just the man to bear the nation's hopes. From the heartland of the Midwest, he had been a Marine combat pilot, won five Distinguished Service Crosses, and married his childhood sweetheart. After nine delays caused by mechanical glitches and bad weather, Glenn was finally strapped into his Mercury capsule on February 20, 1962.

With the rocket billowing icy vapor on the launch pad, with hundreds of thousands crowding nearby beaches and highways, the launch team checked off each system to the flight director. At last the director announced: "All recorders to fast. T minus 18 seconds and counting. Engine start."

"You have a firing signal," Scott Carpenter reported to Glenn through his communicator. "Godspeed, John Glenn."

The blockhouse fell silent as the Atlas trembled, fire pouring from its three engines. Glenn's voice, shaking from the vibration, came through the speakers: "The clock is operating . . . we're under way."

The thousands watching from the streets, the millions watching on television yelled and wept their good wishes as the Atlas rose on its pillar of fire. Inside, Glenn was battered by vibrations and crushed by the acceleration.

Climbing through the sound barrier, still increasing in speed as its fuel load decreased, the Atlas soared upward. The two side-boosters fell away, leaving the third engine to blast Glenn on his way. Finally, the third engine cut off and the Mercury spacecraft, named *Friendship 7*, was on its own. From being six times his weight on Earth, Glenn suddenly found himself weightless.

"Roger. Zero-G, and I feel fine," Glenn's voice came through. "Capsule is turning round. Oh! That view is tremendous!"

Glenn was in orbit, and the watching millions exulted. Knowing that he was addressing the nation, the down-to-earth Glenn did his best to evoke the

1962 John Glenn becomes the first American in orbit
Carpenter lands far off course

May 1963 New Nine join the Original Seven
First woman in space (U.S.S.R.)
Kennedy assassinated

1964 Three men in space (U.S.S.R.)

THE NATION'S HERO

New York had never seen anything like the reception given to John Glenn on February 23, 1962. With each of the Original Seven astronauts in an open car, despite the cold, the motorcade, headed by John Glenn and Vice President Lyndon Johnson, left the airport for lower Manhattan. An estimated 4 million people turned out to scream with delight at Glenn's achievement of being the first American in orbit. There were crowds at the airport, crowds lining the frozen streets and people hanging over the railings above Manhattan's FDR Drive as the seven cars headed downtown.

The motorcade then proceeded at walking pace up Broadway, where the emotion exploded into an intensity that astonished the astronauts. In the canyon of buildings, there were people at every window, screaming, waving and weeping, while at every intersection burly policemen in blue overcoats found themselves crying and yelling along with the crowds. In the tradition of such "ticker-tape" parades, people threw as much shredded paper as they could. Ticker-tape—paper from telexes and stock-market indicators—wasn't all they threw. Broadway was enveloped in a snowstorm of paper—3,500 tons of it.

It was an astonishing expression of the strength of feeling inspired by Cold War rivalry. In this primordial outpouring of emotion, Glenn became the focal point of the nation's fears and hopes. He was, for the moment, the figurehead fashioned from American know-how, patriotism, and good old-fashioned colonial values. It was as if he was a knight who had proved single-handedly that the Soviet dragon could be vanquished and the New Frontier conquered.

glories of his 90-minute day as he peered through his periscope: the blackness of space, the thin band of blue along the sunlit horizon, the snowy mantles on mountains, the deep green of West Indian waters, a string of thunderclouds lit by lightning, the quick spread of hues as he swung into the Earth's dark side, the stars in all their unwinking glory, the splotches and pinpoints of cities in the night below—in Perth, Australia, citizens turned on all their lights for him to see—the dark horizon bursting into bands of reds, yellows and blues as a new day dawned.

Then, suddenly, a mystery. "I'm in a mass of some very small particles that are brilliantly lit up," he said. They were like fireflies, which vanished as he went into darkness and then reappeared in the Sun.

For a moment, on the ground, everyone was riveted by astonishment. Was this some new space hazard? Apparently not: the fireflies, whatever they were, did no damage.

Bad news

Meanwhile, a more serious problem had emerged. The control panel registered a loose heat shield. If this was the case, Glenn could burn up on re-entry. No one told him the bad news. Frantic checks suggested the switch was faulty, not the shield itself. But just in case, engineers suggested a solution. Strapped onto the shield was a set of little rockets that would slow the capsule to start its descent. Normally they would be jettisoned after their job was done. Possibly, if the retropack was retained, its straps would be enough to hold the shield in place long enough to save Glenn's life. No one knew what difference the retropack would make, but it was, perhaps, his only chance. Glenn was told not to jettison the pack.

"Why?" he asked suspiciously.

He was told he would be fully briefed later. Glenn knew an evasion when he heard one, and didn't like the idea. Any change of plan meant a cause-and-effect cascade of potential problems. The retropack would burn off, but might damage the heat shield. Besides, retaining the rockets meant he could not retract his periscope. The open periscope doors would let in heat. When the heat built up, and the pack had burned off, he would have to close the periscope doors. That meant he would have to override the autopilot and fly the capsule manually. If he misjudged, he could bounce back into space or burn up entirely. And if anything else went wrong, no one would ever know, because he would be out of contact, his radio blacked out by the intense heat of re-entry.

Finally, he was told about the shield, and after a moment of anger at the deception accepted the inevitable. After three orbits, four hours into his mission, the retrorockets fired, and at 17,500 mph he fell into the atmosphere. Holding the craft steady with little blasts from the steering rockets, he saw flaming chunks of metal whirl past his window as the retropack burned up. He felt the Gs build, and knew he was slowing.

Below, the ground crew agonized through the 4 minutes and 20 seconds of radio blackout. Then, not knowing if Glenn had survived, Shepard called: *"Friendship 7, this is Mercury Control. How do you read? Over."*

A brief pause, then to everyone's relief: "Loud and clear. How me?"

"Roger, reading you loud and clear. How are you doing?"

"Oh, pretty good."

The laconic right-stuff words released a surge of cheering. America's first orbital astronaut was still falling inside 1½ tons of glowing metal, but he was safe so far, and on target. There was nothing wrong with his heat shield, and never had been. His parachute dropped him into the ocean within a few miles of his rescue vessel.

John Glenn was already a hero, but now the nation's adulation redoubled. After being welcomed by Kennedy, who flew to Cape

SCORCHED RELIC Three days after his safe return to Earth, John Glenn shows President Kennedy and Vice President Johnson (far right) his space capsule, burned on re-entry.

1965 First space walk (U.S.S.R.)
First U.S. space walk by Ed White
Gemini 6 and 7 meet in orbit

1966 Korolev dies on the
operating table
Last of the Gemini missions

Canaveral for the occasion, Glenn addressed a joint session of Congress, as if he were a visiting head of state. He was then whisked to Manhattan, where New Yorkers gave his motorcade, moving at walking pace up Broadway, a "ticker-tape" parade that outdid both Lindbergh's and Shepard's. Glenn was fêted at the United Nations, met the mayor of Perth, who had flown over specially, and then headed to Ohio for an ecstatic local welcome.

Carpenter's close call

The next mission had problems of a different nature. Supposedly, the pilot would be Deke Slayton. But Slayton had a minor heart condition, and NASA's controllers were worried about what the press would say if anything went wrong. Slayton was grounded. His back-up was Walter Schirra. But the next in line, Scott Carpenter, Glenn's back-up, had more simulator time. To the anger of the others—suppressed in public—Carpenter was chosen.

With only ten weeks' preparation for his flight in May, Carpenter then almost brought

disaster on himself. Once aloft, he became so carried away with maneuvering in space that he almost depleted the capsule's fuel supply. Mercury Control nearly brought him home early, then agreed he could stay up if he simply drifted and did nothing. But while floating in his little cabin, his hand struck a wall. At once, a swarm of "fireflies" appeared. Intrigued, he fired his thrusters, turned his craft, and was able to show that the "fireflies" were droplets of human vapor expelled from the craft and frozen onto the sides until knocked off by vibration.

All this took time, fuel and attention. By the time he was due for re-entry, Carpenter was behind on his checklists. He fired his retros three seconds late. That would translate into a major error in landing.

Then, with no fuel left to control his stability, Carpenter's craft began to oscillate wildly, and he had to release his parachutes early. He came down 250 miles from his rescue craft, and out of radio contact. For

BIG BURNER Scott Carpenter, pilot of Mercury 7, America's second orbital mission, flew his space capsule with such gusto that he had no fuel for controlling his descent.

SUIT OF ARMOR Like medieval knights preparing for combat, astronauts needed a team of assistants to get them suited up and in position. This is Gordon Cooper, who flew the last Mercury mission in May 1963.

some 30 minutes, he was entirely lost, which did not look good to a worldwide radio and TV audience.

Although he was acclaimed in public, Carpenter found himself in trouble. There was no evidence in the records of his pulse, but NASA suspected that Carpenter had committed the pilot's cardinal sin: he had panicked. He was never trusted to fly again, and eventually became a Navy aquanaut.

Behind the scenes, the rest of the Seven rallied behind Deke Slayton. Glenn and Shepard came up with a suggestion: if he couldn't fly, why not have him as a boss, as director of Flight Crew Operations in the Astronaut Office? As more astronauts came on line, the group would need someone smart, tough and experienced to select crews and plan training. Within weeks, Slayton had a new role, picking crews for missions he would have given anything to fly on himself.

At once, Slayton won respect and achieved results. In October 1962, he select-

MEDIA HOST Andrian Nikolaev makes the first TV broadcast from space in August 1962. He was joined in space by a second cosmonaut in a separate capsule, making them the first two-man team in space.

ed Wally Schirra to go into orbit. Schirra stayed up for six orbits over nine hours, ran through all his tasks to perfection, yet used up less fuel than his two predecessors.

While the last Mercury mission—Gordon Cooper's 22 orbit flight in May 1963—ran its course, nine more astronauts joined the team. They were dubbed the "New Nine" by *Life* magazine. Their task would be to fly atop the more powerful Titan rocket in two-man missions, code-named Gemini, in order to practice the procedures that would be needed on a Moon mission, such as space walking and docking with another craft.

The Gemini missions would take time to

AMERICA'S LOSS

When he was assassinated in Dallas on November 22, 1963, John Fitzgerald Kennedy had been in office for less than three years. He was 46 years old, and the third president to be assassinated after Abraham Lincoln in 1865.

prepare: in part because no rocket was yet ready to carry the larger two-man capsule; and partly because of the loss of the space project's greatest inspiration. On November 16, President Kennedy visited Cape Canaveral, where von Braun showed off the giant Saturn, ready for another test launch. Six days later, Kennedy was dead.

Khrushchev cranks up the pressure

America's successes were a challenge that Korolev, always under official pressure for more coups, could not ignore. Both nations were now in a race that demanded ever greater achievements: one man in orbit, two men, three men, a rendezvous, space walking, unmanned missions to the Moon and planets, a lunar landing. It became clear in hindsight that for the Soviet Union appearances were as vital as real achievements. Khrushchev needed "firsts" for publicity, and these were delivered like rabbits from hats. The United States, with its efforts always on public show, needed soundly based progress. That difference would favor the Soviet Union in the public mind through much of the 1960s. It would take time for America's commitment to show results.

Khrushchev's response to the American orbital missions was to pre-empt the Gemini missions with a space rendezvous of his own. In August 1962, two R-7s soared up on successive days, placing two Vostok spacecraft into almost identical orbits, less than 3 miles apart. It was a virtuoso achievement of precision rocketry. Headlines in the West spoke of a steppingstone to the Moon. In fact, there was a long way to go, for the Vostoks could not maneuver. Their two pilots, Andrian Nikolaev and Pavel Popovich—Nik and Pop to the headline writers—were little more than passengers. They contributed jokey in-flight conversations, and Nikolaev made the first TV broadcast from space. This was more like propaganda than progress, a minor continuation of Cold War rivalry.

The rivalry took another twist that autumn. Khrushchev had raised the Berlin Wall with impunity; Kennedy had increased American involvement in Vietnam. With few other options available, Khrushchev decided to challenge the United States on the nation's home ground, by installing 40 medium-range weapons in Cuba.

The result was the Cuban Missile Crisis of October 1962. When the rockets were spotted, Kennedy blockaded Cuba. Kennedy's chiefs-of-staff were ready to launch air strikes against the Soviet Union, and to invade Cuba. For several hours, the world seemed on the brink of nuclear war. It was Khrushchev who backed down. He did so, he said, for "the cause of peace." It was certainly in his interests: to go to war would be to lose not only Cuba, but a hefty percentage of his own people.

Space still seemed to be one area where Khrushchev could dominate by pulling off world firsts. The next Soviet achievement was the first woman in space. Valentina Tereshkova had been chosen on Khrushchev's orders to find an ordinary young woman who would show the world that under socialism space flight was for the people. Besides being the right sex, Tereshkova's main qualification was that she was of impeccable Soviet stock—a 26-year-old textile-worker whose tractor-driver father had died fighting the Germans. She was also a terrific skydiver.

SPACE HEROINE Heavily decorated for her achievement, Valentina Tereshkova, the first woman in space, smiles serenely during one of her many public appearances.

Her year of training involved more parachute jumps and some pilot training, but always under close supervision, and it did not give her the expertise acquired by her male colleagues.

When Tereshkova flew into space in June 1963, it was as part of a mission that exactly replicated the previous one. Other than the length of the mission—she was up for three days, and her colleague, Valeri Bykovsky, for five—it had little scientific merit. But it was a significant propaganda coup. Tereshkova became a symbol of Soviet sexual equality, and received star billing on a world tour. That was the fulfillment of Tereshkova's mission. She never flew again, nor did any other Russian woman until 1982. Instead, Tereshkova fell into another role. She and Nikolaev had started an affair before her flight, and the couple got married. With obvious delight, Khrushchev presided over a wedding that was a media extravaganza. Seven months later, Mrs. Nikolaeva had a daughter. The marriage broke up shortly afterward.

Khrushchev wanted more. By early 1963, he had heard of NASA's plans for the two-man Gemini. The U.S.S.R. would have to go one better and place three men in orbit. As it happened, Korolev had a rocket for the job. He had already designed a new ICBM, the R-9, and he now borrowed its upper stage for the R-7 so that it could carry a full 7-ton payload. He also already had a new, better space capsule, named Voskhod ("Sunrise").

But Voskhod was designed to carry just two cosmonauts, not the threesome Khrushchev demanded. Korolev's answer—or rather, the answer of his designer, Konstantin Feoktistov—was to squeeze three men into the two-man Voskhod. The only way to do this was to leave out the space suits and the escape system. As Korolev's deputy, Vasily Mishin, commented afterward, "They were cramped just sitting! Not to mention it was dangerous to fly."

In October 1964, after just four months' training, the threesome went up for 16 orbits.

The operation had concentrated Feoktistov's mind wonderfully, and the three landed safely to further crowing from the Soviet media that America's space program was limping along—a clear indication of capitalism's imminent collapse. In fact, as Mishin said later, the three-man venture was little more than a circus act, "a waste of time." Like previous flights, this one had been Khrushchev's baby, but a day after it returned to Earth, Khrushchev was ejected from the Kremlin. When the three cosmonauts arrived for their official reception in Moscow, they were met by the new leader, Leonid Brezhnev.

Walking in space

The pressure for more space dramas remained. The next step had to be a space walk. Korolev designed an airlock for the Voskhod capsule, and gave his space walker, 30-year-old Lieutenant-Colonel Alexei Leonov, a bulky space suit.

Aloft on March 18, 1965, Leonov eased himself into his tubular airlock, and his partner, Pavel Belyayev, vented the air. Leonov pulled himself out, set up a movie camera, and floated away to the end of his 16-foot tether, the Earth wheeling below him. There was nothing for him to do, so after a few minutes he started to squeeze back in, and found that he would not fit. His space suit had ballooned out and he could hardly bend his limbs. He had to lower his air pressure to a quarter to get back inside. Once there, he collapsed, drained by the effort and tension of his experience.

Yet the stresses of the space walk were merely a prelude to another high drama. After jettisoning the airlock, the two cosmonauts prepared for automatic re-entry—at which point the system failed. They would have to fire their retrorockets manually, one orbit later. But when it entered the atmosphere, the capsule was not aligned properly. It fell too fast, building heat that melted its forward antennae.

Fortunately, the parachutes worked, but the capsule landed 1,200 miles from its intended site without its radio. No one knew what had happened to it. Possibly the pilots were dead. Worse, if they had crashed outside the Soviet Union, the flight could turn into a public relations disaster. A rescue mission swung hurriedly into action.

Leonov and Belyayev found themselves in deep snow, somewhere in the Urals, wedged near the ground between two fir trees, with their parachute caught in branches. Trapped in their freezing capsule, the two men waited for their distress beacon to summon help. No one came. The next morning, a helicopter landed in a clearing, and a rescue unit skied in to lead them out. Only then could the two be whisked back to Moscow for a state reception and the revelation of socialism's newest and greatest achievement. If they had died, no one on the outside would have heard of them until the full details emerged 30 years later. At the time, no one knew how close they had come to disaster, how thin the line was between high-tech safety and death—and for Soviet public relations, between coup and catastrophe.

That was the last of the Voskhod missions. At the time, they had achieved their goals, and the world was convinced that the Soviets were well ahead of the Americans. Yet difficulties would soon beset the Soviet effort.

Soviet burnout

Korolev was about to find that he was being challenged in the field of giant rockets. Vladimir Chelomei, a former designer of submarine-launched cruise missiles, teamed up with Korolev's engine designer, Valentin Glushko. Chelomei wisely employed Khrushchev's son, and got the go-ahead to design a new ICBM driven by six engines that could carry a 30-megaton warhead. Designated the Proton, it would make a superb upper-stage booster, easily powerful enough to carry a man to the Moon.

The rivalry turned into a bitter dispute over engine design, with Korolev rejecting Glushko's ideas. Turning elsewhere for his engines, Korolev came up with support for a Moon rocket of his own, the SL-15, with no fewer than 30 engines, producing a total thrust of 4,430 tons, to Saturn's 3,440 tons. The whole assembly would weigh 2,700 tons, and would carry 95 tons into orbit. But it would take years to test-fly, and would never be a success. In

TESTING FOR THE RIGHT STUFF

From the Original Seven onward, would-be astronauts faced grueling ordeals to prepare them for missions in space. Many of these ordeals derived from pilot training for supersonic fighter-planes, but others were devised simply to subject the men to pressure until every possible source of failure had been eliminated. In pilot parlance, they were being tested to see if they had "the right stuff"—the unflappable calmness that would enable them to perform to perfection, especially if their machines failed.

Probing every muscle, nerve and organ, doctors measured body fat, brain waves and blood composition. The pulse acquired particular significance. It could be monitored all the time, and gave insight into body and mind. A raised heartbeat might mark a man as "panicky" and end his flying ambitions on the spot. For reasons none of the

DESERT TRAINING Two would-be astronauts are pushed to the limits in a survival training exercise. They have made a hut from an old heat shield, and Bedouin-style clothing from space-suit linings.

trainees understood, they were asked to donate sperm so that their sperm count could be measured. There were 17 different eye tests. They were made to perform in water, on treadmills and step machines. They were flown on arcing jet-flights on the "vomit comet," which for a minute or two could create weightlessness. They were dumped in deserts, on mountains, and in the jungles of Panama to practice survival techniques. The regimen was of no practical use to men training for space flight, but it was supposed to reveal and develop character, also assessed by 600-question personality tests.

They were made to endure soundless rooms. They were baked and frozen. They were subjected to rarefied atmosphere and brain-jarring vibrations. Huge centrifuges simulated forces of ascent and descent, squeezing the skin down their faces like putty.

On the whole, the results seemed to justify the demanding procedures. But the men themselves would never have volunteered unless they already had what it took to be an astronaut: intelligence, training and drive.

Little thought was given to what would happen to them afterward. For several, the training and space flights would be the high points of their lives. Readapting to normal life on Earth could prove difficult, and for some it led to bouts of depression and alcoholism.

TIGHT SQUEEZE An astronaut's training was rigorous for a reason: the missions would be arduous, and take place in cramped, nerve-racking conditions, as this picture shows. An astronaut needed to keep a cool head, come what may.

ORBITAL FLIGHTS: THE FIRST FOUR YEARS

Date	Designation (Country)	Astronaut(s)	No. of orbits
1961			
April 12	Vostok 1 (U.S.S.R.)	Gagarin	1
Aug. 6	Vostok 2 (U.S.S.R.)	Titov	17.5
1962			
Feb 20	Mercury 6 (U.S.)	Glenn	3
May 24	Mercury 7 (U.S.)	Carpenter	3
Aug. 11	Vostok 3 (U.S.S.R.)	Nikolaev	64
Aug. 12	Vostok 4 (U.S.S.R.)	Popovich	48
Oct. 3	Mercury 8 (U.S.)	Schirra	6
1963			
May 15	Mercury 9 (U.S.)	Cooper	22
June 14	Vostok 5 (U.S.S.R.)	Bykovsky	81
June 16	Vostok 6 (U.S.S.R.)	Tereshkova	48
1964			
Oct. 12	Voskhod 1 (U.S.S.R.)	Komarov Feoktistov Yegorov	16
1965			
March 18	Voskhod 2 (U.S.S.R.)	Belyayev Leonov	17
March 23	Gemini 3 (U.S.)	Grissom Young	3
June 3	Gemini 4 (U.S.)	McDivitt White	62
Aug. 21	Gemini 5 (U.S.)	Cooper Conrad	120
Dec. 4	Gemini 7 (U.S.)	Borman Lovell	206
Dec. 15	Gemini 6 (U.S.)	Schirra Stafford	15

fact, it could never have been as efficient as Saturn. Korolev's designs assumed that at least one of its engines would fail and the others would have to carry the dead weight. As a consequence, the whole assembly was bigger and heavier than necessary.

Korolev won another round over his rival when he was given control over the Proton's spacecraft, as well as his own Soyuz space capsule. But now, perhaps on the brink of yet further success, Soviet space ventures began to founder. Khrushchev's policy of piling sensation upon sensation, as well as being dangerous and self-deceptive, had squandered time and money that might otherwise have been spent on developing a solid, permanent lead. Korolev's Soyuz spacecraft, which was first launched in April 1967, might well have been in orbit two years earlier, providing a platform for a

Moon mission ahead of the Americans.

As it was, neither Proton nor N-1 were ready, and Korolev's last productive years were wasted. He was a sick man, his health undermined by his years in prison camps. In January 1966, he had an operation to remove an intestinal growth. The surgeon spotted a colonic tumor. He extended the operation, but after eight hours under anesthetia Korolev's heart gave out.

The Gemini missions

Less than a year earlier, the United States had successfully placed two men in orbit aboard Gemini 3. One of those men was to have been Alan Shepard, but he developed an ear problem that affected his balance. Deke Slayton agreed with the doctors: Shepard couldn't fly, at least not right now. To a flying man, this was devastating news. But Slayton had been through the same trauma himself. He offered Shepard an

CHECKING THE HATCHES Gus Grissom and John Young settle into their two-man Gemini capsule. As their mission badge (inset) shows, they named their craft *Molly Brown*, after the 1964 comedy *The Unsinkable Molly Brown*, about a woman who survives the sinking of the *Titanic*.

administrative post and some encouraging words: "I intend to make it into space yet, Al. I'm keeping my foot in the door, and yours should be right there next to mine." They would wait for years, but both would eventually see their dreams come true.

In March 1965, in a capsule 50 percent larger than Mercury, blasted aloft by a Titan booster, Gus Grissom and John Young flew the first manned Gemini mission, Gemini 3. In orbit, they made a few routine in-flight corrections. The only major surprise came when they deployed their enlarged parachute, which brought them up with such a jolt that Grissom smashed his faceplate.

Then, fired by Leonov's space walk, NASA introduced their space suit early so that the next Gemini pair, James McDivitt and Edward White, could undertake their own space walk (or "EVA"—Extra-Vehicular

pushing or tugging, or by firing a small propulsion unit while drifting or twisting around in three dimensions.

Gemini 6 and 7 proved the reliability and accuracy of orbital flying by edging to within a few feet of each other. The pilots of Gemini 7, Frank Borman and Jim Lovell, stayed up a record two weeks, despite being low on fuel and having faulty power cells, in conditions they compared to being locked in a men's lavatory. They returned unshaven, grubby, pale, weak, and with a dramatic loss of weight, but otherwise healthy. NASA was jubilant, for this was proof enough that men

DRESSING FOR SPACE

Space is an environment with unique dangers. Besides airlessness, space walkers face intense solar radiation and occasional dust particles, or micrometeorites, moving at several miles per second. The space suit developed to protect them was an advanced version of the type worn by high-altitude test pilots.

One problem was bulk. If the suit was inflated to normal air pressure of about $14^1/2$ pounds per square inch, it would have to be immensely tough and would become impossibly cumbersome. But if the astronaut is given pure oxygen to breathe, pressure inside the suit can be greatly reduced.

To maintain a steady temperature, the suit must both insulate and control heat. Undergarments contain a network of water-filled tubes that transfer body heat, either into the spacecraft through an umbilical tether or, in later suits, to a backpack. Several layers of Teflon and Fiberglass provide insulation, with enough strength to protect against micrometeorites. Early helmets were close fitting and moved with the astronaut's head. Later helmets were fixed, allowing free head movement within them.

FLOATING ALMOST FREE
Executing the first U.S. space walk, Ed White drifts cheerfully at the end of his tether. When ordered to return to the craft after 20 minutes, he half-joked, "This is the saddest moment of my life."

Activity—in NASA's terminology). With checklists completed, the two let out the cabin's air and opened the hatch. It was White who went out, turning slowly at the end of his tethering-line, controlling himself with a little hand-held gas-jet.

The next mission, Gemini 5, tested radar equipment and new fuel cells over a period of eight days. This was long enough for the astronauts to gain their first experience of eating, sleeping and exercising in space. When they landed, apart from being tired and dehydrated, they were in fine health.

The remaining seven Gemini missions steadily extended the experience of orbital flight. They docked with Atlas-Agena target vehicles, modified orbits, experimented with countless tasks, and built up hours of space walking, learning to cope with snaking umbilical cords, slippery equipment, hooks that could tear a space suit, and entangling wires. All such obstacles and challenges had to be negotiated carefully and patiently by

FOIL WRAPPED Getting suited up was a long process, requiring assistance (left). Once the astronaut put on his gauntlets (above), any delicate work became a nightmare.

LAUNCH DELAYED Wally Schirra leads Thomas Stafford toward their Gemini 6 spacecraft on December 12, 1965—only to have the flight postponed an hour later.

would not be subject to any long-term health risks on a long lunar mission.

In addition to the Gemini missions' countless minor problems, there were also some major ones. Two of the Atlas-Agena rockets exploded, forcing hasty revisions in planning. One launch was aborted after a systems failure, and Gemini 8 was ordered to return early to Earth after a little thruster switched itself on and threw the capsule into a terrifying, vertiginous spin.

Air filters acted up, fuel cells failed, and work in the bulky, unyielding space suits proved far more exhausting than expected. The astronauts found themselves sweating so much that their visors would fog up. Considerable effort was put into developing more sophisticated suits.

Yet there were also no major disasters, and the astronauts brought back astonishing and spectacular film of their work in Earth orbit. The difficulties encountered on space walks inspired new training in a water tank. Re-entry techniques improved, so that several times the parachutes deployed within sight of the rescue ship. Gemini 11, which soared to a record height of 850 miles,

splashed down a mere 1¹/₂ miles from the U.S.S. *Guam*; the astronauts were on board in 24 minutes.

The series ended in November 1966 with Gemini 12. Over 20 months, NASA had logged 1,900 hours of space experience. It had learned many vital lessons which it used to refine both equipment and techniques. In short, NASA had prepared a strong foundation for the giant leap to the Moon.

SPACE RENDEZVOUS After a three-day delay, Gemini 6 took off perfectly (bottom), in time to meet up with Gemini 7 (inset). The two capsules remained a few yards apart for eight hours, paving the way for orbital docking.

TO THE MOON

THE 1960S DEFINED A GENERATION. SET AGAINST A BACKGROUND OF SUPERPOWER RIVALRY, THE SOCIAL CHANGES AND SCIENTIFIC ADVANCES OF THE DECADE TRANSFORMED THE ENTIRE POST-WAR WORLD. IN AN ATMOSPHERE WHEN ANYTHING SEEMED POSSIBLE, PRESIDENT KENNEDY PROMISED TO PLACE MEN ON THE MOON BEFORE THE DECADE WAS OUT, AND THE STAGE WAS SET FOR AN EPIC RACE. THE VICTORS WOULD WIN NOT JUST THE MOON RACE, BUT SYMBOLICALLY THE COLD WAR TOO.

APOLLO—THE EARLY YEARS

AMERICAN SCIENTISTS HAD JUST NINE YEARS TO TURN MISSILES INTO MOON ROCKETS—AND THEY ROSE TO THE CHALLENGE

When the Apollo program was first announced in 1960, no one had a clear idea which rocket would act as the first stage to propel astronauts to the edge of space. The Saturn I, which first flew in October 1961, would be too small. Von Braun, whose Army Ballistic Missile Agency had been reborn as the Marshall Space Flight Center, invited ideas for what he called an "advanced Saturn." Now, for the first time, he was able to look beyond military needs, to his teenage dream of space travel. He reminded the space industry that, in designing the rocket, they should look

SPACE MATERIALS Wernher von Braun, having been transferred from the army to NASA, is shown samples of plastic sandwich material by NASA scientists in 1962.

beyond a manned mission: "It's just a big truck to increase this country's capacity to carry cargo into space."

Boeing was among those who took up the challenge of designing this "big truck." They approached the problem with a practical question: since anything going into Saturn would have to be carried to the site, what was the largest object that could be shipped by road, rail or ship? The answer was around 40 tons. To lift its own weight and a 40-ton cargo into orbit, the advanced Saturn first stage should have about 3,400 tons of thrust. This could be achieved if the Saturn was upgraded with another engine.

Meanwhile, von Braun invited bids for a second stage, and received a massive

proposal from North American Aviation. The documents they submitted formed a stack 2 feet high. North American's design of the second stage, known as S-II, had one great virtue—its light weight. Traditionally, insulation of the super-cold fuel had been on the inside. The S-II's insulation would be on the outside, which meant that the rocket's aluminum shell could be thinner because it would be made stronger by the cold. The weight saving meant that the rocket would provide more than 440 tons of thrust from its engines, yet its shell could be carried on a truck when empty. North American won the contract, making them twice the size of any other NASA supplier.

Just before Christmas 1961, von Braun made his decision. Boeing was given the job

RESEARCH BASE Von Braun's new civilian facility, the Marshall Space Flight Center, operated by NASA, takes shape outside Huntsville, Alabama, in 1960.

of building 24 advanced Saturn boosters, each powerful enough to send a fully-laden railway truck into space. The boosters would be the first stage of a three-stage Moon rocket code-named Saturn V.

Large as Saturn was, people assumed that the assault on the Moon would need two launch vehicles. One would carry the manned lunar spacecraft into Earth orbit. A second would bring up another rocket, which would dock with the lunar craft and enable it to blast to the Moon, land, take off from the Moon, and return to Earth. This approach had no sooner become an accepted

1960

1960 NASA
announces Apollo
program

1961 Boeing awarded
contract to build improved
Saturn boosters

1962 Von Braun accepts
the idea to use only one
rocket per mission

1963 NASA decides to
go with "all-up" testing
to save time

fact than it came under attack, principally from one man, John Houbolt, of NASA's Langley facility.

The standard plan called for a rocket that would have to take enough fuel with it to relaunch from the Moon with three astronauts inside. Houbolt suggested a new idea: lunar-orbit rendezvous. The approach rocket would serve as a mother-ship, sending out a small two-man craft while the third astronaut remained in orbit. It was complex and dangerous. But placing just two men instead of three on the Moon would save 30 percent in weight at the lunar end of the journey, and that translated into a massive saving in the weight to be launched from Earth. Houbolt's idea did away with the need for a second rocket. One Saturn V could launch the whole mission.

Plans are redrawn

The major problem with Houbolt's approach was the level of danger involved for the astronauts. If anything went wrong in an Earth-orbit rendezvous, the astronauts could return to Earth. A disaster in lunar orbit would leave them stranded, facing certain death. Despite the risk, Houbolt's arguments steadily gained ground because the two-rocket approach also had its dangers. In particular, it was impossible to be certain that a Moon rocket launched from Earth orbit would retain enough fuel to leave the Moon.

Steadily, Houbolt won converts. In June 1962, von Braun announced that he, too, had been convinced. In November, when Grumman Aircraft won the contract for the lunar-lander module, Apollo's last major element fell into place.

Sitting on top of three powerful boosters, the complete Apollo spacecraft consisted of three sections. A cone-shaped command module would provide a womb for three astronauts during the six-day out-and-return journey. It would consist of a little room 13 feet across and 11 feet high, big enough for the men to float around freely.

Attached to the rear of the command module was a service module, which would provide power, oxygen, fuel and propulsion. Tucked in behind the service module would be the lunar module, itself in two parts. While one astronaut stayed in lunar orbit in the command-service modules, the whole lunar lander—a weird, spidery thing that needed no aerodynamic molding to land on

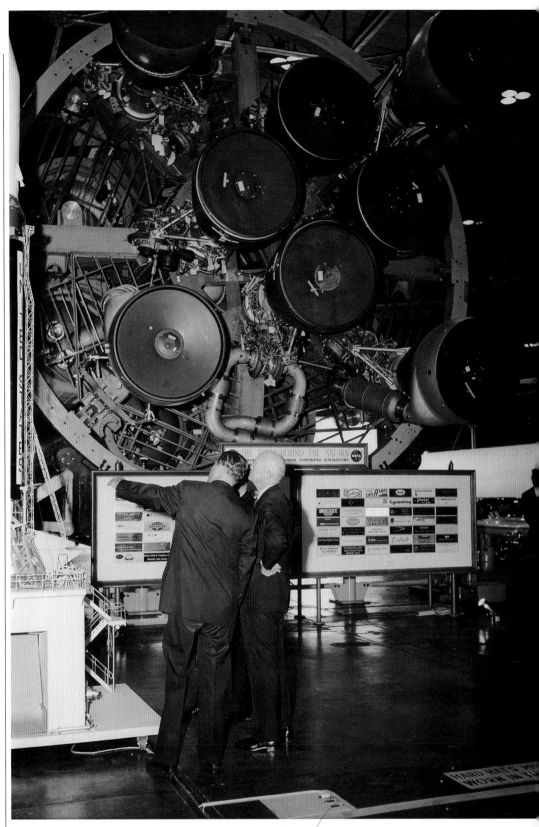

the airless Moon—would descend to the lunar surface. To return, the lander's top section would blast off and rendezvous with the orbiting command-service modules.

The rocket that would set this complex mission on its way would weigh almost 3,000 tons. Almost all this immense weight was

FIRST STEPS TO THE MOON At the opening of the Marshall Space Flight Center in 1960, von Braun (left) briefs President Dwight Eisenhower on the many companies responsible for contributing to the new Saturn I rocket. The test vehicle for the man-on-the-Moon program, Saturn I, would have its first flight in October 1961.

1966 Luna 9 (U.S.S.R.) sends back the first TV images of the Moon
Surveyor 1 (U.S.) transmits 11,000 images of the Moon
Lunar Orbiter 1 (U.S.) provides first overview of the Moon

1967 Apollo 1 command module fire kills three astronauts
Sergei Komarov (U.S.S.R.) is killed in Soyuz 1
First unmanned flight of Saturn V rocket

1968 First manned
Apollo flight—
Apollo 7

fuel, most of which would be used up simply raising the rocket from the Earth. The only item that would return to Earth would be the command module, a mere 0.2 percent of the original weight.

The following year saw NASA make another major time-saving decision. In the past, when rocketry was still in its unreliable youth, all elements were tested in sequence, one at a time. By late 1963, however, a senior NASA administrator, George Mueller, had won a commitment to test the whole rocket all at once, so-called "all-up testing." The decision assumed that the system would work as a whole, leaving only smaller-scale

TRIAL RUN With its eight engines developing 84 tons of thrust each, a Saturn I first stage thunders out in a test firing at the Marshall Space Flight Center in May 1965.

problems to be ironed out. If successful, this approach would save a year or more.

This would be an immense, multi-billion dollar undertaking. The 6 million parts for the Saturn rocket and the Apollo spacecraft had to be collected and assembled from hundreds of subcontractors, and the bits all had to work with unprecedented perfection. Launch failures were a part of life in unmanned rockets. In manned missions, no failure was acceptable. If, on its round trip to the Moon and back, the whole rocket was only 99.99 percent perfect—if it failed by a mere 0.01 percent—that would represent 600 mechanical failures, any one of which might kill the occupants.

Wherever possible, there would be back-up systems. But not everything could be backed up, and any back-up would mean more weight. Every extra pound required an

extra pound of fuel. This, in turn, would mean bigger fuel tanks, more stress on the structure, and more overall risk.

The need for success meant that risks had to be faced. The nation, or rather its leaders, desperately needed good news to counter growing problems. There was no end in sight to the war in Vietnam; casualties were rising, protests growing. Kennedy's commitment to racial equality was foundering under black demands and a white backlash.

Disaster with Apollo 1

In 1966, with the Gemini missions safely behind them, NASA engineers were riding a wave of confidence. Von Braun's Saturn 1-B, a stepping stone to Saturn V, was ready for use. Its second stage would carry the Apollo spacecraft, minus the lunar lander, into orbit the following year, with its three-man crew:

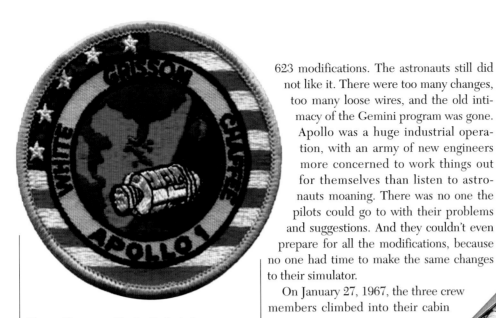

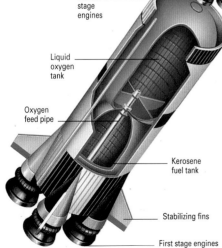

MISSION MEMORIAL The Apollo 1 mission badge for the proposed 1967 launch became a memorial for the three men who died in a ground test in January that year.

Gus Grissom, veteran of Mercury and Gemini, Ed White, who had space walked on Gemini 4, and newcomer Roger Chaffee.

In August, NASA received the first finished command module and gave it extensive tests over the next six months. It was a highly sophisticated creation, designed for lightness with an aluminum alloy held together by 3,000 feet of seams with thousands of welds, some of which had to be accurate to within 0.013 inches. There had been 20,000 failures during the module's construction and assembly, and it was still not up to snuff. Grissom himself suggested countless changes. In January, the module was given

623 modifications. The astronauts still did not like it. There were too many changes, too many loose wires, and the old intimacy of the Gemini program was gone. Apollo was a huge industrial operation, with an army of new engineers more concerned to work things out for themselves than listen to astronauts moaning. There was no one the pilots could go to with their problems and suggestions. And they couldn't even prepare for all the modifications, because no one had time to make the same changes to their simulator.

On January 27, 1967, the three crew members climbed into their cabin on Pad 34 for a full ground test. The rocket was not fueled, and no one foresaw danger. With the hatch sealed, the cabin was pumped full of oxygen. The inside pressure was made higher than normal to prevent air from leaking in.

Monitored on television links by some 200 launch-team members in a concrete bunker known as the Saturn blockhouse, they began to run through the checks to countdown. There were irritating glitches—an odd smell in the oxygen supply, a bit of

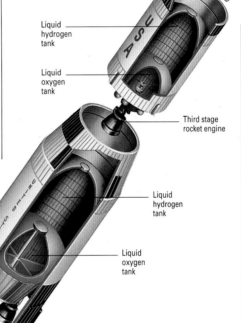

Launch escape rocket

Command module

Service module

Lunar module

Liquid hydrogen tank

Liquid oxygen tank

Third stage rocket engine

Liquid hydrogen tank

Liquid oxygen tank

Second stage engines

Liquid oxygen tank

Oxygen feed pipe

Kerosene fuel tank

Stabilizing fins

First stage engines

SATURN IN PROFILE The giant Saturn V had three main stages. Above these were the lunar module, the service module, and the cone-shaped command module. At the very top was an emergency escape rocket.

crackle in the communications system. After five-and-a-half hours the countdown was nearing an end, with 10 minutes to go, when the launch director called for a hold in order to work on communications.

Eleven minutes later came a quick sound from one of the astronauts: "Fire." Another shouted, "We've got a fire in the cockpit!"

Chaffee was still strapped in, but Grissom dived under a seat and White's arms went up to tackle the hatch bolts. There was nothing they could do. In seconds, flames and smoke

GIANT AMONG ROCKETS

When NASA announced the Moon missions, it said Saturn V "would be longer than a football field, have a base diameter greater than the combined width of three tractor-trailer rigs, weigh more than a light Navy cruiser, and develop more power than a string of Volkswagens from New York to Seattle."

filled the cabin, as controllers watched in horror on the monitors.

Chaffee's frantic voice came through: "We've got a bad fire! We're burning up!"

Then a brief cry, and nothing but a roar. In mere seconds, the cabin had become a

SURVEYING THE MOON

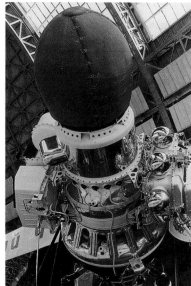

MECHANICAL EGG The Soviet Union's Luna 9 lander contained a TV camera that was protected by a large egg-shaped covering.

As so often in the history of space travel, in the unmanned exploration of the Moon the Soviets were fast starters, the Americans stronger finishers. In the 1950s, Soviet Luna probes had flown past the Moon, crashed onto it, and circled it. In the early 1960s, America started its own close-up research with the Ranger series. The first six failed, but the three later Rangers sent back thousands of detailed pictures of the Moon's surface before they struck it.

In January 1966, the Soviet Union scored another first when Luna 9 soft-landed, and disgorged an egg-shaped object that unfolded and took the first TV shots of the lunar surface. The pictures were blurry, but they showed a porous, pitted, cratered moonscape. Instantly, a mystery had been solved. The Moon was not, as many feared, covered with dust. Spacecraft could land and men could walk there.

Thereafter, America's probes performed with increasing efficiency. They had to, for they had the vital task of identifying a landing site. In June 1966, Surveyor 1 became the first American soft-lander, transmitting 11,000 pictures over six weeks. Five lunar Orbiters followed in 1966-7, and combined to map the surface in detail. Six subsequent Surveyors zeroed in on several possible landing sites. One of them even dug a little trench to give the first analysis of the Moon's crust (it was basalt, evidence of volcanic activity in the distant past). For the first time, scientists also had a detailed picture of the Moon's hidden far side.

By February 1968, the work was complete. The next American visitor to the Moon would be a human one.

SEARCHING FOR SITES Surveyor 1's first shots of the lunar surface in June 1966 (below) included its own foot, seen at the bottom of the photograph. In August 1966, Lunar Orbiter 1 (above right) gave the first detailed overviews, which included the Moon's far side (right).

2,500 degree oven, producing such pressure that in the 16th second the walls ruptured with a whoosh, blasting out dense black smoke, the heat driving back the five launch pad personnel. The astronauts were still alive at this point. They were killed when their oxygen hoses burned through and carbon monoxide blasted into their lungs, asphyxiating them. They were unconscious after half a minute, dead within two.

When rescuers could finally get inside, they found the corpses in their blackened suits covered in congealed electrical insulation and nylon netting.

As the word spread, it fell to a young rookie, Alan Bean, to coordinate the release of the news to the dead men's wives. In the neat ranch houses of El Lago, where astronauts, engineers and administrators clung to suburban normalcy, other wives gathered to be with Pat White, Martha Chaffee and Betty Grissom.

The subsequent investigation confirmed the two main causes of the disaster: the gas and the hatch. Oxygen is notoriously dangerous as a fire hazard. But there was no source of ignition that anyone knew of—until a spark suddenly flew from some hidden electrical connection. As for the hatch, there was no quick-escape system. There had been one on the earlier Mercury flights, until Grissom's capsule almost sank when its hatch unaccountably blew open after splashdown. In the early Apollo capsule, this could not happen. There were two hatches, the inner one fastened by six bolts, which had to be loosened with a wrench. And the hatch opened inward. Even in perfect conditions, it would take 90 seconds to get out. With the pressure inside the capsule already above normal, and rising at an explosive speed, the hatch was pinned shut by gas pressure alone. White had had no chance of opening it.

At a deeper level, the disaster revealed an underlying malaise produced by haste and inexperience. A detailed analysis of Apollo's sister-craft being made by North American Aviation revealed over 1,400 faults. Apollo 1 had been a disaster waiting to happen.

There was, however, no thought of giving up. The pilots themselves knew the dangers, and Grissom himself had said in one of his many press conferences: "If we die, we want people to accept it. We're in a risky business." He could not possibly have guessed that the most dangerous place was not rising into orbit on top of 2,000 tons of flaming fuel, nor inside a plummeting capsule, but on the ground, in an unfueled rocket surrounded by technicians.

It would take half a billion dollars and almost two years of reorganization, redesigning and rebuilding to iron out the problems raised by those terrible few seconds.

The Soviet "devil machine"

Three months later, in Central Asia, Sergei Korolev's successor, Vasily Mishin, set in motion the first manned Soyuz mission. There had been four unmanned tests, with unpromising results. One or more major problems had beset each flight. But like their American counterparts, the Soviets were in a hurry. Though the details were suppressed at the time, it emerged later that the administrators tried to short-circuit a step-by-step approach: one-man, two-man, link-up, space walk. Instead, they gambled on a double mission. First up would be Vladimir Komarov, who had commanded the cramped three-man Voskhod mission in 1964. Then would come a three-man launch, a docking, a transfer of two cosmonauts, and the return of both craft together.

Disaster struck early, when one of Komarov's solar panels failed to deploy. There were no back-up batteries. Then his main radio transmitter failed. Finally, the automatic control system malfunctioned, sending the craft tumbling. "Devil machine!" Komarov swore. "Nothing I lay my hands on works!" He switched to manual, and stabilized his craft. But without enough power, the mission was doomed. The second rocket remained on its stand, and Komarov prepared for re-entry after just one day.

On the way down, his main parachute failed to open, and his reserve parachute got tangled in it. There was no escape; like the Apollo, the Soyuz capsule had no ejector seat. Beneath its useless "Roman candle" of twisted parachutes, Komarov's craft crashed onto the steppe at 400 mph, too fast to give his retrorockets time to cushion the impact. The craft split apart and exploded. Komarov died instantly.

His death had much the same effect on the Soviet space program as America's tragedy, except that the responses remained secret. A few weeks after the funeral, Yuri Gagarin, in a long and officially sanctioned newspaper interview, declared: "The road to the stars is steep and dangerous. But we are not afraid . . . Space flights cannot be stopped. This isn't the work of any one man or even a group of men. It is a historical process." In a further cruel twist of fate, Gagarin would himself become a victim of this same process when his training plane crashed the following year.

Back in business

For another year, both sides worked frantically to get their damaged programs back on track. In the United States, NASA administrators and politicians fought over the costs and benefits of the Apollo program. Was it really worth $25 billion to send a man to the Moon? The Cold War seemed to be winding down. The boom of the early 1960s was over. Taxes were rising. Polls showed that public opinion was not as firmly behind the space program as it had been. The country was convulsed by two linked crises: the civil rights movement and the Vietnam War.

Kennedy's successor, President Lyndon Johnson, shocked the nation by declining to stand for re-election in 1968. But neverthe-

FALLEN HERO The widow of Vladimir Komarov mourns her husband at the Kremlin Wall, where his ashes were given a place of honor. Seen in his space suit (inset), Komarov had commanded one of the most dangerous Soviet missions in 1964.

less Johnson wanted action, and he wanted it fast. Tangible results were needed for the sake of the Democratic Party, for the sake of

his Southern associates and their companies, and for the sake of what Kennedy had promised. There was not much time left.

The first giant Saturn V, 363 feet tall, flew unmanned in November 1967, ascending on a volcanic pillar of fire visible 150 miles away. All three stages performed perfectly, delivering into orbit a working command-and-service module and a dummy lunar module weighing over 120 tons—more than the combined weight of all the previous 350 American satellites added together. The following January, the lunar module had its first trials in orbit.

A second Saturn V test, in April 1968, hit a snag. The first stage had begun to vibrate so intensely that two engines shut down early. Then another engine refused to start.

It turned out that a fuel line had snapped. Yet back on Earth, identical copies of the lines tested perfectly.

The makers, Rocketdyne, were baffled, until they realized that on Earth the subzero fuel caused ice to form, which held the lines rigid. In space, there was no ice, so the lines were free to vibrate—and eventually break. Once identified, the fault was easy to fix. But there was still no lunar module, which was in trouble with faulty wiring and a temperamental lunar ascent rocket.

Meanwhile, the Soviets were once again confident of progress. In September, a variation of Soyuz called Zond looped around the Moon and returned to Earth. Korolev's new SL-15 lay almost ready in Tyuratam. It was surely only a matter of time before a Russian cosmonaut headed for the Moon.

Apollo 7, the first manned Apollo mission, was due up in October, carried aloft by a Saturn I-B. Apollo 8, set for early 1969, was due to test the Saturn V and command-service module together—everything but the lunar lander. But even if everything went right, there seemed little chance of meeting Kennedy's deadline. Then George Low, a

APOLLO IN ORBIT The expended upper stage of the Saturn I-B rocket falls away from the Apollo 7 spacecraft, leaving it safely in orbit on October 11, 1968.

Houston-based engineer who oversaw the development of the spacecraft, suggested a daring shortcut: why not send Apollo 8 around the Moon? They could practice lunar orbit and test the craft at the same time.

It made sense—if Apollo 7 was a success.

On Apollo 7, the only problem was a human one. The commander, Wally Schirra, had a cold which made him a misery to his colleagues and a pain to his controllers. He canceled a scheduled TV broadcast, only to reinstate it, under pressure. He called his controllers idiots for imposing what he called Mickey Mouse tasks. But otherwise, all went well.

The Soviets, too, were back in business. Two weeks after the Apollo 7 splashdown, Soyuz 2 and 3, each with a single cosmonaut inside, practiced a successful close approach. The next month, another automatic Zond probe circled the Moon and returned safely. But now NASA was ready, morale restored by Apollo 7's successful low-key flight. The nightmare of the Apollo 1 fire behind them, they were ready to shoot for the Moon.

APOLLO PIONEERS The crew of Apollo 7— (left to right) Donn Eisele, Walter Schirra and Walter Cunningham—pose for a publicity shot after being chosen to crew the first manned Apollo mission. Each is wearing the mission badge (inset).

1968—A TIME OF TRANSITION

THE SPACE RACE, WAR, ASSASSINATION, DISSENSION AND REPRESSION—THESE ALL MADE 1968 A DEFINING MOMENT FOR AN ENTIRE GENERATION

While NASA struggled to fulfill Kennedy's promise to put a man on the Moon by the end of 1969, the country was beset by a succession of catastrophes, both national and international.

In January, North Vietnamese forces stormed the American Embassy in Saigon during the Tet ("New Year") offensive, showing that the world's greatest power could not, after all, batter a nation of peasants into submission. President Lyndon Johnson, Kennedy's successor, declared he would not run again for the presidency, in effect abandoning any attempt to solve the Vietnam issue.

Civil rights remained another wound that divided the nation. In April, Martin Luther King, Jr., the inspiring leader of the civil rights movement, was gunned down on a motel balcony in Memphis. Within hours of his death, there were race riots in dozens of cities.

Simultaneously, other parts of the Western world seemed on the verge of anarchy, while the Soviet Union could apparently assert itself at will. Student unrest, bringing with it fear of revolution, spread in Germany, Holland and Britain. In Paris, rioting students seized the center of the city while a series of strikes paralyzed all of France.

In Czechoslovakia, Alexander Dubcek's regime raised hopes that liberalization would spread through the Soviet empire, until, in August, Soviet tanks rolled into Prague to shatter that illusion. Although both sides in the Cold War were talking about limiting nuclear arms, the Soviet Union proved itself as ruthless as ever within its own empire.

At home, John Kennedy's mantle seemed to settle on the shoulders of his younger brother, Robert, who announced that he would seek the Democratic nomination for the presidency. If anyone could realize the twin dreams of escaping from the nightmare of the Vietnam War and bringing equality to a racially divided society, surely it was Robert F. Kennedy. But in June, he, too, fell to an assassin's bullet.

SHATTERED DREAMS A two-mule wagon and 50,000 people formed the funeral cortege for Martin Luther King, Jr., on April 9, 1961. On August 9, in Prague (below), a defiant young Czech waves a flag from atop a Soviet tank—part of the force that crushed Dubcek's "Prague Spring."

RIDING THE FIRE

THE SECOND MAJOR STAGE OF PREPARING FOR A MOON LANDING TOOK APOLLO ASTRONAUTS FROM EARTH ORBIT TO LUNAR ORBIT

After an intensive debriefing of the Apollo 7 crew, it was agreed that all systems were as ready as they would ever be. On November 11, 1968, President Lyndon Johnson gave his blessing for Apollo 8, America's single greatest gamble in the race to put a man on the Moon.

The three crew members—Borman, Lovell and Anders—had known this was in the cards since the summer. Frank Borman had a reputation for bluntness, stubbornness, and instant but well-founded decisions. He had supervised the redesign of the command module after the Apollo 1 fire.

Jim Lovell's sunny optimism was a good match for Borman's gruffness. They were both 40 years old and had worked well together on Gemini 7. The third crewman was Bill Anders. He and Neil Armstrong had been in training that spring to fly the lunar lander. So when he learned of his mission with Borman, his delight was mingled with disappointment. He would go to the Moon, but he would probably never land on it.

The three had four months in which to train for the mission. They were scheduled to fly on December 21, the first to ride the Saturn V into space, testing every piece of testable equipment. If all went well they would, in NASA's phrase, go for "trans-lunar injection" (TLI)—in plain English, they would relight Saturn's third stage, accelerate

SLOW START At Cape Canaveral, a huge crawler transporter, traveling at a mere 2 mph, takes Apollo 8 out to its launch pad in October 1968.

GETTING READY Left to right: Frank Borman, Jim Lovell and Bill Anders in a mock-up of the command module during training for their Apollo 8 flight.

out of Earth orbit, cast off the third-stage booster and head for the Moon.

This would be something wholly new in manned space flight. Gemini 11 had set the record by going out to 850 miles. Apollo 8 would travel almost 300 times that distance—234,000 miles out into space. Accuracy in direction and speed was vital. The Moon orbits the Earth at 2,300 mph, and they had to fall into its gravitational field without approaching too close. They would fire the main engine to slow the module to 3,700 mph, allowing it to enter lunar orbit. Their flight was timed to get them the best

---

HEAVY PRESSURE

When a Saturn V rocket lifted off from Cape Canaveral, it created pressure waves that were sensed in the Lamont-Doherty Geological Observatory in New Jersey— 1,100 miles away.

---

view of the Sea of Tranquillity, a possible landing site, in the sharp, low light of the rising Sun. After ten orbits, they would fire the engine again to accelerate homeward. Finally, they would eject the service module and prepare the command module for re-entry, traveling at 25,000 mph. Their flight path was critical: more than 1 degree too shallow, and they would skip off into space like a flat stone skimming water; more than

1 degree too steep, and they would burn up.

All this was to be practiced on Earth in simulators that would mimic both perfect performance and a mass of malfunctions. Would it all work in practice? The engineers said Apollo was "three nines" reliable, meaning 99.9 percent. In mechanical terms, that was unprecedented, but was it good enough? There had never been a manned mission so far into space before, so there were no real-life statistics or comparisons.

A historic visitor

On December 10, the three astronauts retired to the simple crew quarters, where they would live in quarantined isolation, making final preparations for the launch 11 days later. Eight miles away, the Saturn towered above Pad 39A.

On the day before their launch, they had a visitor. Nothing, and no one else, could have given them such a sense of the significance of their journey. It was Charles Lindbergh, the man who had first flown solo across the Atlantic Ocean 41 years earlier. He described how he had worked out his fuel requirements by using a piece of string to measure the distance from New York to Paris on a library globe. Lindbergh did a quick mental calculation and then said: "In the first second of your flight tomorrow,

FIRE AND THUNDER Apollo 8 lifts off on December 21, 1968. It would leave Earth orbit on a figure-eight swing around the Moon—as shown on its mission badge (inset).

you'll burn ten times more fuel than I did all the way to Paris."

The next morning, the astronauts shared a traditional breakfast of steak and eggs before climbing into their white, fireproof suits. The bubble helmets clicked on, the oxygen flowed and they headed stiffly for the bus that would take them to Pad 39A, where the spotlit Saturn stood breathing fumes of frozen vapor. An elevator ride, a walk over a gantry, and they nestled into their couches. Technicians swung the hatch shut.

With just under 2 1/2 hours to go, they began to run through the familiar litany of checks. Below them, 1,000 tons of super-cooled liquid fuel flooded the tanks. Pressure built. Heart-rates climbed. The last checks, the last switches set. Nine, eight, seven . . . Fuel began to gush from the engines . . . Ignition! Smoke and flames, then white fire tinged with gold. The rocket began to shudder, building power, struggling

BORMAN LOVELL ANDERS

THE DISCOVERY OF EARTH

The Apollo 8 astronauts were so busy that they were on their fourth orbit of the Moon before they noticed a sight that touched their souls. As Borman turned the craft so that Lovell could take a sextant measurement of their position, Anders glanced up and exclaimed: "Oh, my God. Look at that picture over there."

Beyond the washed-out desolation of the Moon's surface, above the horizon, there was a marbled semicircle of blue and white.

"Hey, don't take that, it's not scheduled," said Borman, with typically unemotional practicality.

Anders ignored him. "Hand me that roll of color quick would you?" But Lovell was staring at the glorious sight. "Oh, man, that's great!"

"Hurry. Quick." Anders got his film, loaded the camera and aimed the telephoto lens. "You got it?" asked Lovell. "Take several of them."

There was a brief pause, and Anders had the shot that captured their first heart-stopping sight of the Earth—the vivid blue of the oceans speckled and streamed with the brilliant white of clouds, through which showed the oranges and tans of deserts, and the surprising grey of forests—green does not penetrate the atmosphere well—all forming a marbled glory that outshone the stars.

The photograph would become a classic, reproduced in countless forms in books, calendars and posters. It was the first time people had seen the Earth as a whole, in color. Above any other image, it seemed to capture something unexpected about space exploration. Framed against the utter black of space, set against the bleakness of a dead world, the view made millions feel at a glance that they had glimpsed a new and startling insight into their own smallness, fragility and uniqueness. In Anders's words: "We came all the way to the Moon, and the most important thing is that we discovered the Earth."

"EARTHRISE" This memorable image shows how normal ideas of up and down do not apply in space. The Earth's South Pole, and Antarctica, are on the left side, while the edge of the shadow runs north-south through Africa.

to escape its restraining bolts . . . Three, two, one . . . Liftoff!

The Saturn raised itself uncertainly. It seemed tortured by spasms. This was something new, something not programmed into the simulators. Below them, the giant engines sensed every minute deviation and corrected with mighty jolts, raising the rocket slowly. It took 10 long seconds to clear the launch tower, but acceleration was building as the great ship leaned and blasted itself into its flight path on a pillar of fire.

After 40 seconds, Saturn went supersonic and the roar died. From Houston came the voice of communications officer Mike Collins: "Apollo you're looking good." Still it accelerated as the burden of fuel fell away.

Three times the force of gravity, four—then a sudden release from pressure as the first stage shut down. A muffled bang. The emergency escape rocket blew away, uncovering the windows. The second stage fired.

Houston again: "Apollo 8, your trajectory and guidance are go." At 8 minutes 45 seconds, the second stage fell away and the third stage kicked in with a small jolt. Then, just 11½ minutes after launch, silence. Apollo 8 was 115 miles up, moving at 17,400 mph, in a perfect orbit around the Earth.

They had almost two orbits, or 2½ hours, for checks. This was the time for the final decision. If there was anything wrong, there would be no "trans-lunar injection," no lunar mission, and almost certainly no man on the

Moon before the decade was out. Minutes passed, while the hundreds of engineers in Houston checked and rechecked their instruments. Aboard Apollo, all was in order.

Finally, Mike Collins spoke the words that would send the first men beyond the Earth's gravity: "Apollo 8, you are go for TLI."

To the Moon

With 10 seconds to go, the computer flashed a request for confirmation. Lovell pushed a button marked "Proceed." The third stage fired again, gently building speed, driving them up and away from Earth. After exactly 5 minutes and 18 seconds, the engine shut down, having accelerated them by another 7,000 mph. Collins' voice came through:

"We have a whole room full of people who say you look good."

Borman cut loose from the third stage, fired the service module's little thrusters, and swung them around. There was the third stage looming large, and there, fitting neatly into the round porthole, was the Earth. This was the first time human eyes had seen the whole of it, a jewel set against the blackness of space. But there was no time for anything except work. The third stage dwindled away behind them until Apollo 8 was alone, turning slowly once an hour to balance the heat from the Sun and the cold of space. The astronauts called it "barbecue mode."

A few hours out, they could relax, climb out of their suits and float in zero gravity. Out here, there was a new physics. Objects

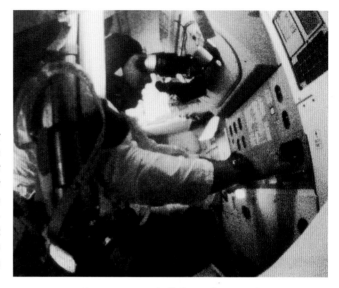

CHECKING POSITIONS Like an ancient mariner, Jim Lovell takes sightings from the stars to check the spacecraft's position.

drifted and rolled lazily, water formed trembling balls. And the command module was roomy compared to Gemini. Here, they could move around. For some astronauts, the sensation some-times induced nausea, but that soon passed. In deep space, zero gravity gave the same delight as in orbit.

Eighteen hours into the mission, Borman awoke from 5 hours of fitful sleep, assailed by flu-like symptoms. He threw up into a bag, but not accurately enough to prevent a

pulsating green ball from floating free and hitting Lovell in the chest. Then Borman went down with diarrhea. Bits of vomit and feces flew slowly round the cabin, until they were hunted down with paper towels. The stench was appalling, a problem they shared with Houston on a special closed telemetry channel when Borman recovered. Not that it would have made a difference if he had been really sick; they were too far into the mission to abort now. The only way they could get back was to go around the Moon.

More than halfway, the crew turned on their TV camera. The pictures, picked up by giant radio dishes and transmitted to Houston via a landline, were blurry black-and-white images, but good enough to show wonders: space technology, the strangeness of this weightless environment, and surpris-ingly ordinary things made extraordinary by the context. Anders twirled his toothbrush. Lovell wished his mother happy birthday. "Jim is fixing dessert," commented Borman. "He's making up a bag of chocolate pudding. You can see it come floating by." The Earth, he said, was very, very beautiful. People had to take his word for it; the camera couldn't handle the contrast, and showed only a blob of light. They couldn't see the Moon, either. It was lost in the glare of the Sun beyond.

All this time Apollo had been slowing as it climbed the gravitational hill away from the Earth. Finally, two days and 7 hours out, 207,000 miles from Earth, and 38,900 miles from the Moon, traveling at a mere 2,223 mph, they topped the hill and began to "fall"

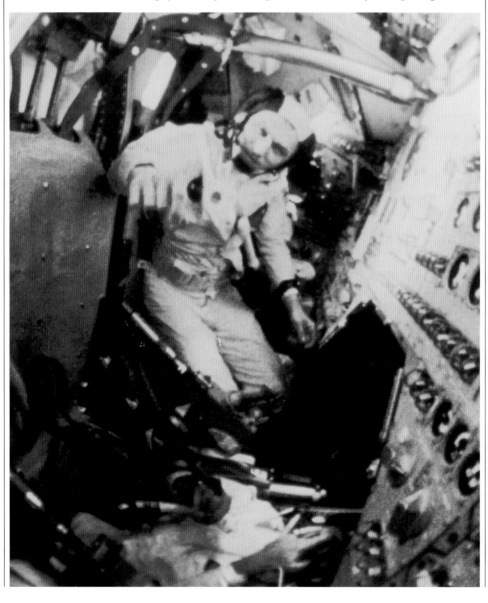

CLOSE QUARTERS The Apollo command module was spacious by comparison with earlier spacecraft, but still required care when moving around in zero gravity.

NEW VIEW On their way back home, the Apollo 8 crew was granted a stunning view of the Moon as it had never been seen before. On the right-hand side is part of the Moon's far side, here illuminated by the Sun.

toward the Moon. They could not certify this for themselves; they knew only because Mission Control told them: "You're going for LOI (lunar orbit insertion). You're riding the best bird we can find."

The engine had to fire for exactly the right time—247 seconds—to drop the craft into its orbital low-point, 69 miles above the Moon's surface. Since they were flying on their backs to absorb the pressure of the engine firing, they couldn't see where they were going. Their only guide was from Houston, where the computer knew to the second when they would experience "loss of signal" as they went out of radio contact behind the Moon.

"Ten seconds to LOS."

"Thanks a lot, troops," said Anders.

"We'll see you on the other side," Lovell added cheerfully.

At 68 hours, 58 minutes and 4 seconds, Apollo 8 vanished from Houston's radio link. Accelerated by lunar gravity, the craft was now traveling at 5,000 mph, 1,300 mph too fast. Houston would not know until Apollo re-emerged whether the craft had slowed into lunar orbit or not. If not, it would re-emerge sooner, whipping around the Moon on its way back home: that was the insurance in the event of engine failure.

Inside Apollo, the light suddenly failed. They were in a vast shadow. Anders craned his neck and for the first time saw the stars in their multitudes, and a giant black hole where there were no stars. It was the first human glimpse of the far side of the Moon. Seven minutes later, they burst into sunlight, though not into radio contact with the still-hidden Earth.

Lovell glanced up. "Hey, I got the Moon."

Anders looked up, and saw what looked like oil slicks streaming by on the window. Then he refocused his eyes, and realized that he was looking at lunar mountain ranges flowing past below.

"Oh, my God," he said.

"What's wrong?" Borman was anxious.

"Look at that."

"All right, all right, come on. You're going to look at that for a long time."

The engine fired. They heard nothing, and felt only a vibration and the pressure on their backs as weight returned. The vibration went on, and on.

"Jesus, four minutes?" said Borman.

"Longest four minutes I ever spent," Lovell commented.

The engine cut. They were weightless again, and right on course, accurate to less than a mile, in a mildly elliptical orbit. Later, there would be another small correction to make the orbit circular, a regular 69 miles from the surface.

Now they could take in the view. It was one of utter barrenness, a monochrome of grays. Clear as a close-up, they saw craters of all shapes and sizes, smaller ones overlapping larger ones, massive ones ringed by bleak and pockmarked mountains.

Minutes later they heard Houston calling: "Apollo 8 . . . Apollo 8 . . . Apollo 8 . . . "

Suddenly, there was Borman's calm voice: "Go ahead, Houston."

Those three words turned Mission Control into a bedlam of cheering and whistling. Communications captain (capcom) Jerry Carr asked what the Moon looked like.

"Okay, Houston," said Lovell. "The Moon is essentially gray. No color. Looks like plaster of Paris . . . "

"Or a beach," Anders suggested.

" . . . or a sort of grayish beach sand."

There was work to be done, pictures to be taken. Anders, the photographer, had two Hasselblads and a 16mm movie camera to record assigned targets. But the windows were smeary and small, and it was hard to know where he was. When he was finally able to look up, on the fourth orbit, he saw the Earth in an astonishing setting that moved him, and would later move millions: "earthrise" over the Moon's desolate horizon.

It was Lovell's job to scout for a possible landing site in the Sea of Tranquillity. Drifting above, he searched for obstacles, and saw none. He declared that it looked like a great place to land.

After 14 hours in orbit, they were tired. Borman had slept, and now he ordered the other two to do the same, even if it meant cutting into the planned tasks. Nothing was more important than retaining alertness for

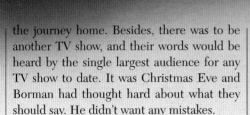

the journey home. Besides, there was to be another TV show, and their words would be heard by the single largest audience for any TV show to date. It was Christmas Eve and Borman had thought hard about what they should say. He didn't want any mistakes.

As Earth rose again on their ninth orbit, Borman went on air to half-a-billion people around the world, with words appropriate for an audience that was predominantly Western and Christian. "This is Apollo 8, coming to you live from the Moon," he began. Then, after each man summarized his reactions, the three of them in turn read from the opening of Genesis, beginning with Anders: "In the beginning, God created the heaven and the Earth . . . " Finally, with the black shadows lengthening on the Moon below, the blue-and-white Earth low on the horizon, and radio silence approaching, Borman concluded: "'. . . and God saw that it was good.' And from the crew of the Apollo 8, we close with good night, good luck, a Merry Christmas, and God bless all of you, all of you on the good Earth."

They vanished behind the Moon for the last time. The engine ignited on schedule and burned for 304 seconds. They reappeared, accelerating out of orbit for the 2½-day return journey. "Please be advised there is a Santa Claus," said Lovell to Houston. "The burn was good."

On the way back to Earth, the astronauts found a surprise in the food locker: real turkey, cranberry sauce and three mini-bottles of brandy. The bottles went unopened— Borman hated to think of the reactions by the media if anything went wrong.

Re-entry would be done by the computer, with Borman piloting only in case of a malfunction. Falling toward the Earth, accelerated by gravity to 25,000 mph, Borman ditched the service module, leaving it to turn into a meteor above the Pacific. Then they were dipping and soaring as the heat shield burned away, the drogue parachutes deployed, the main chutes cracked out, and

the chatter of the rescue helicopters filled their headsets. They were coming down 25 miles from their ship, the U.S.S. *Yorktown.* "Welcome home, gentlemen, we'll have you aboard in no time."

On splashdown, wind whipped the capsule over before Borman could release the chutes. They were upside down, pitching and tossing enough to make Borman throw up. Then three balloons inflated and turned them right side up, and swimmers secured a flotation collar. The hatch opened, and suddenly, in the inrush of fresh salt air, they became aware of the acrid smell that had built up in their week-long journey.

Testing all systems

The achievements of Apollo 8 gave the United States a clear lead in the race to the Moon. The goal that had been set by Kennedy was within reach, but there were several vital stages to go: space-testing the lunar module, testing the moonwalk space

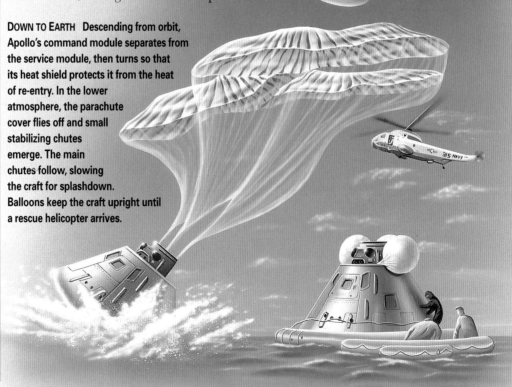

DOWN TO EARTH Descending from orbit, Apollo's command module separates from the service module, then turns so that its heat shield protects it from the heat of re-entry. In the lower atmosphere, the parachute cover flies off and small stabilizing chutes emerge. The main chutes follow, slowing the craft for splashdown. Balloons keep the craft upright until a rescue helicopter arrives.

suit, and the rendezvous between lander and command module. There would be just two chances to do all this: with Apollo 9 in Earth orbit, and Apollo 10 in lunar orbit. Then would come the landing itself.

With the plans laid, the crews emerged from the roster of names. Just into the new year, 1969, Neil Armstrong, Buzz Aldrin and Mike Collins learned they would fly Apollo 11 to make the first lunar landing.

But the Soviets were not yet prepared to concede. In January 1969, Soyuz 4 and 5 met in orbit to make the first docking of two manned spacecraft. After the two craft had linked together, two cosmonauts donned space suits, climbed outside and pulled themselves along handrails from one craft to the other.

The mission was billed as a vital step in Soviet lunar landing plans, but it was in many ways a dead end. No one would ever transfer between manned craft externally. The parade that followed, the rejoicing and the televised award ceremony were largely for show. The Soviets had achieved something that the Americans had not yet attempted, but their program was not as

MEETING UP IN SPACE

Docking in space is a tricky, topsy-turvy operation. As if they were marbles in a funnel, the slower moving of two objects falls into a lower orbit—and then, because its orbit is smaller, it overtakes the upper, faster object. If you slow down, you seem to go faster, and vice versa.

If you want to catch up, you must first slow down and drop into a lower orbit. Only when you are almost level with your target do you accelerate, rising to match your target's height and speed.

In December 1965, Gemini 6 and 7 had maneuvered to within a few feet of one another, but Soyuz 4 and 5, illustrated here, were the first two manned craft to dock in space.

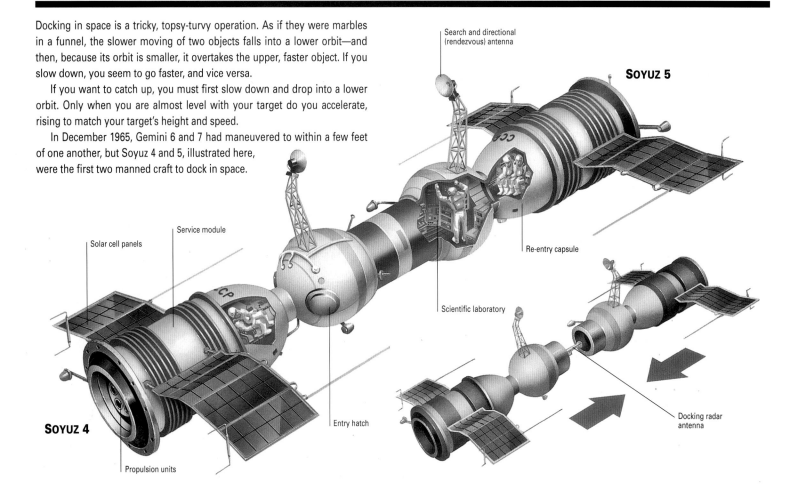

Search and directional (rendezvous) antenna

SOYUZ 5

Re-entry capsule

Scientific laboratory

Service module

Solar cell panels

Docking radar antenna

Entry hatch

SOYUZ 4

Propulsion units

well-focused. Besides, they had a major problem to contend with.

The Soviets had discovered that their heavy-lift booster, the SL-15, could not be brought on-line fast enough. Gambling in an attempt to match the Saturn V's success, Mishin fired his first SL-15 in February 1969. This immense machine, 30 percent more powerful than Saturn, and capable of lifting 4,500 tons, failed a minute after take-off. Vibrations ruptured a fuel line, a fire broke out, engines exploded, and several thousand tons of flaming metal and fuel plunged onto the steppe. As Mishin said, such a failure was not unusual for a first launch. But this was a failure of particular importance: it meant there was now no chance of beating the Americans to the Moon, unless Apollo itself failed.

Final test runs

In fact, Apollo 9, in March, was a test-pilot's dream. Although it stuck to Earth orbit, the mission was more dangerous than the ride around the Moon. First the team separated the service-command module from the rocket's third stage, then turned around and docked with the angular, insect-like lunar module, *Spider*. The lunar module was the first vehicle ever designed for use solely in space. It had a skin so thin it bulged when pressurized, and its spindly legs gave it an odd, organic look.

Two pilots—Rusty Schweickart and Jim McDivitt—crossed from the command module, and took the lunar module on a 111-mile trial run. Then, dancing with the forces that governed spacecraft as they changed speed and orbit, they brought it carefully back. All went well.

The next mission, Apollo 10, was a dress rehearsal for the real thing. But why not go all the way? One reason was that Apollo 10's lunar lander was heavier than the final version. Another was that no one was quite sure how the Moon's lumpy, irregular gravity might affect a low-flying lander. There were simply too many unknowns.

The crew were veterans of five

Gemini flights in all, and knew each other well. The commander, Tom Stafford, had known the command module pilot, John Young, for 20 years, and had commanded Gemini 9 with the lunar module pilot Gene Cernan. Their flight, in May, followed Apollo 8's course, except that they had their lunar module, nicknamed *Snoopy*, attached to the nose of their command module, *Charlie Brown*.

Once in lunar orbit, Stafford and Cernan left Young to orbit in the command module while they took *Snoopy* down to 47,600 feet—only 9 miles from the surface—to take a close look at two possible landing sites in the Sea of Tranquillity. Traveling at 3,700 mph, they seemed to skim the surface. For the first time, they got a true sense of the ruggedness of the lunar terrain. Mountain ranges loomed, rimming impact craters that spanned many miles across. There were boulders ten stories high. Then the plains of Tranquillity came into view, and the favored landing site. It looked smooth, though there were some boulders nearby.

Now came a moment of truth. Their fate lay with one little engine that would blast them back up to *Charlie Brown*. They cut loose from the descent stage, sending it crashing onto the Moon, then fired the ascent stage. It worked perfectly. The way was now clear for Apollo 11.

THE GIANT LEAP

THE FIRST HUMAN STEPS ON THE MOON WERE WITNESSED LIVE BY ONE-SIXTH OF THE EARTH'S POPULATION

At the time of Apollo 11's launch, there was a degree of complacency in the watching world. It was assumed that NASA would not risk lives, so the rocket would not take off unless it was going to work perfectly.

This complacency underestimated the risks involved. The hardware was as reliable as it could be, but things could still go wrong for any number of reasons. The whole enterprise could fail or succeed according to the character and skills of its crew.

The Apollo 11 team mustered a powerful range of talents. The commander was Neil Armstrong, whose quiet reticence concealed a rigorous intellect. He had calmly steadied Gemini 8 when it started tumbling. In the bedstead-like lunar landing training vehicle, he had mastered the technique of balancing the lander on its single exhaust jet.

Edwin Aldrin was the second lunar module pilot. He was known from childhood as Buzz because that was how his baby sister pronounced "brother." Aldrin had a Ph.D. in astronautics; no one knew more about space rendezvous than he did. Michael Collins, the skilled command module pilot, provided the easy, good-humored responses that members of the press sometimes found hard to elicit from the other two.

Even in the best circumstances, the three men would face formidable challenges, in particular during their Moon landing. The lander's computer, working a rocket engine with a throttle, could control deceleration and descent to a few hundred feet. It could even take the lander right to the surface. But no sensor could reveal the perfect site. The computer might land them in a crater or a boulder-field. It would be up to the pilot, Armstrong, to realign the lander onto a new,

NATIONAL HEROES Apollo 11's crew—(left to right) Armstrong, Collins and Aldrin—pose before their great adventure. Their mission badge shows their landing craft as a bald eagle, the national emblem.

better site. If something went wrong, it would be up to Armstrong to press the "Abort" button and fire the ascent stage. There would be no second chance.

Armstrong was not the only one who needed the skill and decisiveness to override the computer. In lunar orbit, Collins might have to implement any one of 18 different emergency rendezvous maneuvers. In Houston, Eugene Krantz, a former Air Force pilot, presided over a 20-man team that monitored every system on the descending lunar module. Each controller had his own team of experts to draw on. From all the simulated landings they had done, from all the conceivable disasters that they had analyzed, they could consult libraries of print-outs to identify faults and suggest responses.

Journey to the Sea of Tranquillity

For three days after their launch on July 16, the astronauts restrained their anxieties and performed routine tasks as they traveled toward the Moon. Four days out, in lunar orbit, Armstrong and Aldrin transferred to the lunar module, named *Eagle*, leaving Collins in the command module, *Columbia*. Back in Mission Control, Krantz and his team monitored everything, passing comments through communications captain Charlie Duke, while reporters listened over an open communications loop. There was another loop for private conversations.

After separation, *Columbia* re-emerged from its first swing behind the Moon. *Eagle* appeared 2 minutes later, descending backwards. At 40,000 feet, the landing radar started to feed speed and altitude data to the computer. As the surface came closer, crisis loomed. A warning light came on, and the data read-outs kept cutting out and cutting back in again. The computer kept flashing an alarm numbered "1202."

In Houston, Krantz was baffled. There were so many possible faults no one could remember them all. He asked 26-year-old computer expert Steve Bales if "1202" would force an abort. No, came the reply, a flashing "1202" meant the computer couldn't read out its information quickly enough. But

UP AND AWAY Apollo 11 blasts off from Cape Canaveral on July 16, 1969—a little over five months short of the deadline set by President Kennedy eight years earlier.

OVER THE MOON The Moon forms a backdrop to the orbiting command module, from which Armstrong and Aldrin have just separated in their lunar module, *Eagle*.

there was no malfunction. Only if the warning was steady did it mean trouble.

"We're Go on that," said Bales firmly. A second later, Krantz passed on the decision: "We're Go on that alarm."

Armstrong knew the Sea of Tranquillity well from simulator training, and it was clearly lit. Takeoff from Earth had been timed so that the Sun would be 10 degrees above the horizon to give the best lighting. He was aiming for a flat area some 11 miles by 3 1/2 miles in size—at least it looked flat from 9 miles up.

At 7,500 feet, with the long deceleration over, *Eagle* was descending at a stately 50 mph, and drifting forward at the same speed. At 1,000 feet, Armstrong saw that the chosen landing area was a field of boulders. He overrode the computer, accelerated to cut the descent, and flew on, looking for a clear spot. Aldrin read out the essentials: height, descent speed, horizontal speed, fuel. In Houston, they knew that Armstrong had control, but not why. There was nothing they could do but keep quiet and wait.

"300 feet, down three-and-a-half, 47 forward," came Aldrin's deadpan voice.

"How's the fuel?"

"Eight percent."

"Okay," said Armstrong. "Looks like a good area here."

"220 feet, 13 forward . . . 11 forward."

That area was not right either: they were over a crater. More valuable seconds had been used up, leaving less and less fuel. They had 1 minute and 10 seconds of fuel left. Then, 200 feet ahead, was a smooth apron of dust. Armstrong brought *Eagle* to a halt, standing vertically on its exhaust. It descended slowly, sending up a haze of dust.

Charlie Duke's voice came from Houston: "Sixty seconds." But 20 seconds had to be kept for a possible abort. They had 40 seconds of fuel with which to land. In Houston, with everyone riveted to their monitors, no one spoke. *Eagle* shifted back and forth, and Armstrong corrected. Dust blew wildly.

"Thirty seconds."

"Contact light," said Aldrin, as the long, delicate probes beneath the landing pads touched the lunar surface. *Eagle* settled so gently there was no sense of impact.

"Shutdown," said Armstrong, and Aldrin ran off a quick post-landing checklist.

There was a moment of silence between Moon and Earth. Then, at 3:17 p.m. Houston time, Armstrong spoke, deadpan as ever: "Houston, Tranquillity Base here. The *Eagle* has landed."

They were safe, with 20 seconds to spare. The news took 1 1/2 seconds to reach Houston. Another 1 1/2 seconds later came Duke's voice, clearly full of relief. "Roger, Tranquillity, we copy you on the ground. You got a bunch of guys about to turn blue. We're breathing again. Thanks a lot."

Stepping out onto the Moon

Eagle was on a level plain pockmarked with small craters and scattered rocks. In the middle distance, ridges stood out with unnatural clarity in the vacuum. No one knew yet exactly where they had landed, but *Eagle* was in perfect shape. There was no

urgency. There were checks to be made, and they had even planned to sleep, if necessary, before stepping onto the Moon.

However, both men were ready and eager to proceed. They had long felt the significance of what was happening, knowing that they were about to mark a turning point in history, not simply for their nation, but for all of humanity. Armstrong knew that his first step onto the Moon would be over quickly, and had a few pithy words ready in his mind.

But first came the arduous business of dressing, including the life-support system with its breathing apparatus and water that would circulate through their suits, and an outer helmet with a gold-plated visor to

---

" THE LONELIEST OF MEN?

Michael Collins, on his own in the command module, describes his reactions to being out of radio contact above the Moon's dark side:

"Far from feeling lonely or abandoned, I feel very much part of what is taking place on the lunar surface . . . I don't mean to deny a feeling of solitude . . . I am alone now, truly alone, and absolutely isolated from any known life. I am it. If a count were taken, the score would be three billion plus two over on the other side of the Moon, and one plus God knows what on this side. I feel this powerfully—not as fear or loneliness—but as awareness, anticipation, satisfaction, confidence, almost exultation. I like the feeling. Outside my window, I can see stars, and that is all. Where I know the Moon to be, there is simply a black void. To compare the sensation with something terrestrial, perhaps being alone in a skiff in the middle of the Pacific on a pitch-black night would most nearly approximate my situation . . . [But] I am cut off from human voices for only 48 minutes out of each two hours . . . I feel simultaneously closer to, and farther away from Houston than I would if I were on some remote spot on Earth, which would deny me conversation with other humans for months on end." "

---

reflect the Sun's fierce glare. On Earth, their gear would have weighed 360 pounds each. Here, it was only one-sixth that weight. At last, they let out *Eagle*'s oxygen, and Armstrong backed out of the hatch.

The world, and a live television audience of about 600 million people—one-sixth of humanity—was waiting. Armstrong's first act was to release a small television camera, which started transmitting a low-angle shot of his descent. *Eagle* had landed so gently that it rested on the surface dust instead of sinking in, so the ladder did not reach the

As seen by a small TV camera mounted on the *Eagle,* Armstrong prepares to make his historic "small step" onto the dusty surface of the Moon.

ground. Armstrong jumped, and dropped in slow motion onto the footpad at the base of the lander. He made a brief comment about the fine-grained surface, then said: "Okay. I'm going to step off the LM now."

His left foot went forward. He flexed a knee, making sure he was firm, and said: "That's one small step for man; one giant leap for mankind."

The sentiment was just right, asserting the minuteness of his movement, and the immensity of its significance.

Armstrong moved around in stiff strides, but with a floating motion familiar from training simulations on Earth. He gathered a bagful of dark, powdery dust and small

REFLECTED GLORY Monitored by mission control in Houston (above), Buzz Aldrin poses for the camera on the Moon. In his visor are distorted reflections of his shadow, the *Eagle* lander and photographer Neil Armstrong.

1970

July 16, 1969 Apollo 11 blasts off for the Moon

July 20 Neil Armstrong and Buzz Aldrin become the first men to stand on the Moon

July 21 Armstrong and Aldrin leave the lunar surface after 21 hours, 36 minutes

July 22 Service-command module sets off for Earth

July 24 Splashdown in the Pacific Ocean

TRANQUILLITY BASE With the American flag unfurled in an imaginary lunar breeze (left), Aldrin sets up a seismometer to record moonquakes, and mirrors to assist scientists in measuring the precise distance to the Moon. Having completed their work, the astronauts blasted off in the *Eagle*'s ascent stage (above), here photographed by Collins as it approached the command module.

rocks, took pictures, and studied the land-scape. It had, he said, a stark beauty. He raised the television camera on a 60-foot pole to give viewers an overview.

After Aldrin emerged into what he termed this "magnificent desolation," the two set up an American flag stiffened with wire. Though this act recalled the way in which past colonists claimed possession of new lands, there was no such intention, for an international treaty of 1967 ruled that no nation could make territorial claims in space.

Then Aldrin, testing movement, did what seemed a dance of joy, leaping, twirling and bunny-hopping in slow motion, leaning unnaturally far forward to counter the mass of his pack, leaving splashes of dust and trails of sharply detailed boot prints.

There was also work to do: more rocks to gather, pictures to take, two experiments to be set up. In the first, a seismometer would detect moonquakes; in the second, mirrors would reflect a laser shot from Earth. By timing the flash, scientists could measure the distance to the Moon to within one foot. Armstrong measured a nearby crater and

gathered more rocks from farther afield. After 2½ hours, their time was up. They climbed back on board, slept fitfully, and then prepared for the critical blast-off by the lunar module's little ascent engine.

Going home

For the 21½ hours that his colleagues were on the Moon, Collins had been circling above, nervously imagining the nightmare that would follow if the ascent engine failed and he was left to go home alone. However, in the eye of the camera that the astronauts left behind, the ascent module vanished from *Eagle*'s legs in a puff of smoke and a spray of debris. Inside, the ascent felt like 7 minutes in a high-speed elevator. From above, Collins saw the dot that was *Eagle* climbing toward him.

The two modules maneuvered closer, and docked. After the astronauts had transferred themselves and their two caskets of rocks to the command module, they cleaned off the penetrating, fine lunar dust as best they could; it smelled acrid, gunpowdery. Then the ascent module was jettisoned.

Early on July 22, the service module engine sprang to life and the three were bound for home—but not yet to the sort of reception that had greeted previous space travelers. To minimize the remote risk of spreading some lunar microbe unknown on Earth, after splashdown on July 24 the men were sealed into isolation garments and placed inside a quarantine trailer. Debriefing took place during two weeks spent isolated

CHECKING FOR MOON BUGS

Apart from the usual dangers of fault and error, there were things about the mission that no one could know in advance. In particular, NASA had to take seriously the idea that the Moon sustained life. It was hard to think of a more sterile environment, but no one was going to risk the remote possibility of a lunar microbe sticking to an astronaut's boot and then escaping on Earth.

A first clean-up was done in lunar orbit, after Armstrong and Aldrin had entered the command module. The astronauts vacuumed themselves, then allowed the air in the module to be sucked into *Eagle,* which they then dumped in space.

Back home, NASA had devised "biological isolation garments," which were tossed into the capsule after it splashed down in the Pacific, as if the occupants were plague victims. From their helicopter, the astronauts entered a special quarantine trailer, which isolated them during their trip back to Texas. Then they spent two weeks in the Lunar Receiving Laboratory. In this sealed apartment, where the facilities included a bar and an exercise room, they waited while scientists watched anxiously for new and terrifying symptoms to appear.

Nothing happened; no lunar bugs were found. The Moon was as sterile as it had always seemed.

CAREFUL WELCOME Using a microphone, President Nixon greets the cheerful Apollo crewmen in their special quarantine trailer aboard the rescue vessel, the U.S.S. *Hornet.*

in the Lunar Receiving Laboratory.

When the three men finally emerged, they were whisked to Los Angeles for a presidential banquet. To some, it seemed they were celebrating not simply the fulfillment of Kennedy's promise, but the end of an era. Others saw the mission as a beginning.

THE END OF APOLLO

THE LATER APOLLOS HAD A NEAR DISASTER AND MANY SUCCESSES— BUT THE MOON RACE WAS WON, AND APOLLO'S DAY WAS OVER

Apollo 11's success inspired grand ambitions for space travel: 100 men in Earth orbit, a manned lunar orbiter, a manned Mars mission, all within ten years or so, and all this beyond the nine additional Moon missions that were planned for Apollo. But these ambitions were already meeting resistance. NASA's budget was being curbed even as Apollo fulfilled Kennedy's promise.

Kennedy was long dead, his dream had been realized, and the passions of the Cold War were petering out in negotiation.

Apollo, though, had a momentum of its own, especially since there were already nine more Saturn Vs on the assembly line. With no further strategic or political role to fulfill, it was decided that these missions would concentrate on the scientific prob-

lems raised by the first landing, researching canyons and craters that would throw light on the Moon's evolution, and possibly the Earth's as well.

Apollo 12 would land on a vast lava plain known as the Ocean of Storms, which might provide rocks that were younger and chemically different from those already recovered from the Sea of Tranquillity. Moreover, there was a target: Surveyor 3, the probe that had landed there in April 1967. To aim at this tiny bull's-eye would test the skills of Mission Control and the astronauts to new limits.

It was hard to see how such accuracy was

MOON VETERAN Pete Conrad prepares to remove the TV camera from Surveyor 3, a probe that had landed over two years earlier. Behind him stands Apollo 12's lunar lander.

1970 Apollo 13 comes close to tragedy
Luna 16 (U.S.S.R.) collects a soil sample by robot
Luna 17 (U.S.S.R.) lands the Lunokhod 1 robot vehicle

1971 Apollo 14 lands on the lunar uplands
Apollo 15 explores the Apennine uplands in Lunar Rover

1972 Apollo 16 lands in the Descartes highlands
Apollo 17 discovers orange soil with glass droplets

possible. No one yet knew how the Moon's "lumpy" gravity affected the orbiting craft, or how to counteract it. Then a young mathematician in Mission Control pointed out that a change in gravitational pull would result in a minute but measurable change of speed, and therefore of frequency, in radio waves. This would allow the orbit to be recalculated accurately enough to guide the descending spacecraft to a pinpoint landing. Then, while Dick Gordon remained in lunar orbit, the commander, Pete Conrad, and Alan Bean would spend 31 hours moonwalking, ending by retrieving part of Surveyor 3.

Lightning strikes

November 14, 1969, was a rough day at Cape Canaveral. As Apollo 12's countdown proceeded, the rocket was lashed by rain from cloudy skies. Takeoff was perfect. But then, as the rocket tilted to aim for orbit, there was a sudden burst of light, a jolt, static filled the intercom, alarms rang, and every warning light flashed on.

Conrad called Houston: "I don't know what happened here. We had everything in the world drop out." In Mission Control, the flight director wondered whether to abort. But the rocket was still on course, and the second stage kicked in normally.

Conrad guessed what had happened. As the rocket lifted, Saturn had trailed a column of gas, creating one vastly extended lightning rod. Thirty-six seconds after takeoff, a bolt of lightning struck the rocket, and traveled down to the launch tower. The surge had caused the command module to shut down. But nothing had affected the engines themselves, and the circuits revived. There was relief after this stark reminder of the unforeseeable risks of space flight.

After a trouble-free journey, Pete Conrad took the lunar module down toward the Moon. He'd never believed the computer could guide them to an accuracy of a few feet, but when the module set itself upright for the final descent, he saw the pattern of craters he was looking for, picked out by the low, early morning light. "Son of a gun!" he exclaimed. "Right down the middle of the road!" He assumed control, maneuvered to find a flat spot, and landed in a flurry of rocket-blown dust. Conrad gave his first step a touch of humor in tune with his 5-foot, 6-

FEELING HEAVY After ten days in space, the Earth's gravity can come as a shock. Here a Navy diver helps Alan Bean climb out of Apollo 12's command module, bobbing on the Pacific Ocean near Samoa.

inch frame and reputation for joking. "Whoopee! Man, that may have been a small one for Neil, but it's a long one for me."

It was a big step for science as well. The instruments planted in the lunar soil by the two astronauts would measure moonquakes, the magnetic field, the incredibly tenuous atmosphere, and the Sun's "wind" of high-energy particles. The only instrument that failed was the television camera: it burned out when Bean pointed it directly at the Sun. But the other odd-shaped instruments began work instantly. Back home, scientists saw a receiver screen wiggle as the seismometer recorded the astronauts' footfalls.

After sleeping, the astronauts took a second moonwalk in the lunar dawn—since each day on the Moon lasts two Earth weeks, their 13 hours back in the module was the equivalent of a lunar half-hour. For the second excursion the astronauts became geolo-

AGONIZING TENSION As Fred Haise and his two partners fight for survival aboard the crippled Apollo 13, Haise's pregnant wife Mary seeks reassurance from a NASA official (above). Meanwhile, in Mission Control (below), six astronauts and two flight controllers make calculations, monitor the progress of the stricken craft, and offer tense advice to its crew. The line-up includes Alan Shepard (seated, second from right), the first American astronaut and soon to become commander of Apollo 14.

gists, collecting as many different rocks as possible and making notes of where they found them. They reached the spindly Surveyor lander, and snipped off its camera for the people back home to see what 31 months of standing on the Moon had done to it. Then, with a perfect mission behind them, they headed for home, and an uncertain future for the rest of the Apollo project.

Unlucky Friday the 13th

Although Apollo 12 went without a hitch, the lightning strike had been a reminder that space exploration was a high-risk business. In January 1970, budget cut-backs led to the cancellation of Apollo 20. Apollos 18 and 19 were in the balance, and some people were questioning whether the remaining missions were really worth the risks and costs involved. Those fears were nearly borne out by Apollo 13.

Two days into the mission, on Friday April 13, 1970, the crew—commander Jim Lovell, Jack Swigert and Fred Haise—had just finished a tele-

cast, when Mission Control told Swigert to switch on fans that would stir up the super-cold liquid hydrogen and oxygen in the service module's fuel cells; these supplied electrical power as a by-product of making water.

Seconds later the crew was startled by a dull bang. The craft shuddered. A warning light indicated a loss of power to the command module, named *Odyssey*. Haise, who was in the lunar module, *Aquarius*, pulled himself back into *Odyssey*. The whole ship had been blown into a slow tumble by the force of the explosion.

"Okay, Houston, we've had a problem," said Swigert, in a classic example of right-stuff understatement. He sounded so calm that the communications captain, or "capcom," Jack Lousma, took a second to respond. "This is Houston. Say again please."

This time Lovell came in. "Houston, we've had a problem. We've had a main B-Bus undervolt." Lovell had traveled farther than any other

human, as a veteran of Geminis 7 and 8 and Apollo 8. At age 42, this would be his last flight. He had never seen anything like this.

Haise, the electrical expert, checked the

NEW AGE LIFESAVER

Apollo 13's lunar module, *Aquarius*, was named after the Age of Aquarius, an astronomical phase opening over the next millennium. Its beginning is defined as the time when precession—the 26,000 year cycle of change in the Earth's axis of rotation—moves the spring equinox from the zodiacal constellation of Pisces into that of Aquarius. Astrologers claim that this will mark a New Age of universal awareness. The term "Age of Aquarius" became a catch phrase in the 1960s, especially after it was popularized in song in the musical *Hair*.

power to "Bus B," a junction point leading from the third of the three fuel cells. He found that the No. 3 cell was dead. Mission rules stated that all three cells, on which all the equipment depended for power, had to be working for the craft to go into lunar orbit. They had just lost the mission.

But there was more. As Haise reconnected the failing systems to Bus A, which linked fuel cells 1 and 2, he found that the No. 1 cell was dead too. Only No. 2 was working. Worse still: of the two oxygen tanks that fed the fuel cells, one was empty, the other reading one-third normal pressure.

It took months of work later to establish what had happened. Two weeks before the mission, engineers had experienced trouble emptying oxygen from tank No. 2. In the end, they forced the oxygen out by heating it, but the safety switch controlling the heater failed, allowing heat to build until it melted the insulation on a couple of wires. When Swigert threw the switch to start the fan, a spark leaped between the two naked wires, ignited the oxygen, and blew the tank open. That severed pipes to the second tank, shut down the fuel cells, and closed the lines to the thrusters. The venting gas acted as a jet, spinning the ship. The thrusters and engine were out, the fuel cells dead, the oxygen tanks dying. In 105 minutes, they were going to run out of power and air, and they couldn't even stabilize the craft.

These multiple, simultaneous failures were so unlikely that no one had thought of simulating them. It was the mechanical equivalent of a healthy man suffering a stroke, a heart attack and suffocation all at once. Moreover, they were past the point of no return. There was no electricity to fire *Odyssey*'s engines, which could have allowed them to make a U-turn. Worse, they were not on a "free-return" course—a path that, in the event of failure, would allow them to swing around the Moon and slingshot a return to Earth. On its present course, Apollo 13 would indeed whip around the Moon, but would miss the Earth by many thousands of miles.

As *Odyssey*'s systems died, the crew spent precious minutes figuring out the only possible solution. They would use the lunar module, *Aquarius*, as a life raft, correcting their course and speed with its little engine.

In normal circumstances, *Aquarius* would have taken 2 hours to power up. They had minutes. And they also had to transfer vital information about their position, speed and alignment to *Aquarius*'s computer. Only then would it be possible to work out the direction and length of firing necessary to correct

around the Moon, and back to Earth. But that was just the beginning. *Aquarius* was designed to support two men for the 45-hour round trip to the Moon's surface and back into orbit. Now it would have to supply the oxygen, power and water for three men for twice that long.

There was oxygen enough in the lander, the lunar backpacks and reserve bottles, but the electricity supply was a problem. The lander had no fuel cells to make electricity, only batteries. These would last only if as many things as possible were turned off, reducing power consumption to the equivalent of a couple of refrigerators.

A third problem was lack of water, vital both for drinking and as a coolant for the electrical supplies. The crew's supply, 338 pounds, would leave them 5 hours short. But they knew, from experiments, that *Aquarius* would probably last another few hours without water. By rationing themselves, they guessed they could make it, but only if a sec-

their course. To do this, Swigert had to draw on precious power from *Aquarius*'s batteries, which would be vital for re-entry. Once *Aquarius* was powered up, all three crowded in and then they shut down the damaged service-command module. Lovell turned his attention to fighting the venting gas from the service module with *Aquarius*'s jets.

About 6 hours after the accident, *Aquarius*'s engine rumbled into life. The first of four perfect computer-controlled burns put them on a course that would take them

GROUNDED EXPERT Ken Mattingly (left), barred from Apollo 13 a day before the launch with suspected German measles, provides life-saving information on how best to keep the command module working.

ond burn succeeded in accelerating them and cutting their return journey from four days to 2½. The crew should have returned in *Odyssey* alone. *Aquarius*, with less power, had to do all the work with the dead weight of *Odyssey* attached.

HAPPY LANDINGS Jim Lovell, commander of Apollo 13, reads a newspaper report of his own safe return. At this point, Lovell and his companions were aboard the recovery vessel U.S.S. *Iwo Jima*, heading for Hawaii.

carry them through the atmosphere to splashdown in the Pacific. The latter required a checklist of procedures that would have taken three months to write and verify on Earth. But Houston had an ace up its sleeve: the talents of Ken Mattingly, an astronaut who had been bumped from Apollo 13 at the last minute because he seemed about to develop German measles. No one knew the command module better. The accident jolted him out of depression into an intense effort to provide his colleagues with the information they needed.

Another major concern now came to the fore: exhaled carbon dioxide. This was normally cleaned from the air by canisters of chemical "scrubbers." But the supply in the lunar module was only enough for two men for two days. There were other canisters in the command module, but they wouldn't fit on *Aquarius*'s piping. Within a day, the men would begin to suffocate on their own exhaled breath. NASA engineers worked out a way to improvise an air purifier from odds and ends they knew were on board: a book cover, plastic bags, sticky tape, a sock.

With only the radio and the inadequate environmental control system switched on, *Aquarius* was using a mere 12 amps. The temperature dropped, and damp air produced condensation. The

No one had ever foreseen such a weird circumstance. It meant that Lovell would have to fire the engines by hand, before the computer took over.

Around the shadowed far side of the

BRAVERY AWARD In Hawaii, President Nixon congratulates Lovell during a ceremony in which all three astronauts were awarded the Presidential Medal of Freedom, America's highest civilian honor.

Moon they swung, just 136 miles from the lunar surface, and out again into the sunlight. That evening, almost 24 hours after the accident, Lovell pushed the "Start" button for the engine that should have been landing him on the Moon, and slid the throttle open. The computer took over to time the burst.

It worked. A minute later, they were at least on a course that would take them straight back to Earth.

Now the problems were twofold: how to endure, and how to revive the moribund *Odyssey* needed to

ROUGH RIDE INTO ORBIT

Alan Shepard, writing in the third person in *Moonshot*, describes the experience of liftoff in Apollo 14:

"The astronauts felt a gentle sense of motion, mildly jerky at first, but to their surprise 'a very gentle rise...'

"Ten seconds passed after lift-off before the first-stage engines cleared the launch tower. A ponderous, slogging, slow-motion beginning, but not for long. Thirty seconds passed... Now with telling effect G-forces increased as the giant continued to accelerate. Saturn V slammed into the area of maximum aerodynamic pressure—that jagged reef in the heavens where sonic waves hammered and pounded against the great body, trying to rip inside... Shock waves formed like ghostly dervishes dancing along the circular flanks of the rocket. A mist appeared and expanded upward above the howling engines: ionized gas, shock waves, plasma in maddening motion.

"Inside... the speed of sound now far behind, it was eerily quiet. Had the crew not heard the humming of electronic equipment in the command module, they might have been in a simulator on the ground...

"'Stand by for the train wreck,' the astronauts called.

"Two and a half minutes from first motion, G-forces made them weigh four times more than they did at lift-off. The five great engines of the first stage had compressed the entire rocket like an accordion until first-stage shut-down. Without constant acceleration and with sudden cut-off of stage one, the three men jerked forward in their seats. The accordion stretched out and compressed again. The fuels sloshed, and the astronauts felt a series of bumps just like a train wreck.

"They heard metallic bangs and assorted noises as explosives separated the now empty stage. They were nearly 40 miles high ... climbing faster then 6,000 mph."

astronauts began to refer to the command module, which they used as sleeping quarters, as "the refrigerator." While the world waited, the astronauts snatched fitful sleep, zipped into sleeping-bags anchored at odd angles. But the cold was intense. They all shivered constantly, rubbing hands and feet to keep their circulation going.

Early on Thursday—six days after the accident—Ken Mattingly was ready with the checklist that would, he hoped, revive the dormant *Odyssey* and prepare it for re-entry. It took him some 2 hours to dictate the sequence, switch by switch.

The next day, conditions in *Aquarius* were like those in a damp cellar in winter. Panels and windows misted, quivering globules of water floated about and clung to pipes. The temperature was in the mid-40s Fahrenheit. Conditions were foul in other respects as well. The men, dehydrated, frozen and exhausted, had been ordered not

to dump their urine overboard, in case the exit-pipe froze. Instead, they had to collect their urine in their space suit bags, which they kept attached. Now Haise was developing a urinary tract infection.

As they approached Earth, the crew was told to cast off the service module, leaving only *Aquarius* and the command module, *Odyssey*, attached. As the stricken service module floated away, they saw for the first time the full extent of the damage. The explosion had disemboweled her. Tanks and pipes stood out like entrails.

Two-and-a-half hours before re-entry, the time came to revive *Odyssey*, wiping off mist-covered gauges and throwing switches to send an electric current along dampened wires. There were no short circuits. The three men crowded back into the command module and cast off their lifeboat, *Aquarius*. With a jolt, the spidery craft drifted away to a fiery death in Earth's upper atmosphere. "Farewell *Aquarius*," said capcom Joe Kerwin from Houston. "And we thank you."

On Earth, millions waited in an agony of suspense through the 4-minute radio silence imposed by the heat of re-entry. When the parachutes burst open, swinging Apollo 13 to a perfect splashdown, the cheering rang around the world.

The final missions

Almost instantly, Apollo 13 became the supreme space adventure. NASA still refers to it in its official literature as the "successful

failure." Nothing could have demonstrated better the combination of skill, heroism and scientific endeavor that characterized the Apollo program. Nothing could have shown to better advantage NASA's wealth of intellectual and human resourcefulness. But not even all of this could revive the spirit kindled by Kennedy almost ten years before. The national mood was focused on Vietnam, not space adventures. In the end, the nation's leaders decided that Americans did not need manned space missions after all, and could not afford them. Only by the narrowest of margins did Congress approve the funds for four more Apollo missions.

NASA needed a new rationale, and found it in science. This was something that had never been foreseen in the decade leading up to the start of the missions. Yet now it was science that set the agenda, inspiring longer spacewalks, better backpacks and more complex data-gathering devices. Lunar cars would take the astronauts ever farther in their expeditions.

In early 1971, on Apollo 14, Alan Shepard became the only Mercury astronaut to get to the Moon. It was a personal triumph. After being grounded by a disorder of the inner ear, he had administered the Astronaut *continued on page 98*

DEEP HEAT The later missions concentrated on scientific discovery. Here, Apollo 15's David Scott drills a hole to bury sensors that would measure heat from the Moon's interior.

FROM "BURNT POTATOES" TO A NEW MOON

On the evening of July 25, 1969, five geologists in white, hospital-style clothing stood around the vacuum chamber in the Lunar Receiving Laboratory at the Manned Space Flight Center in Houston. Inside the chamber was a silvery box, and inside the box were pieces of the Moon. Here, unseen by anyone on Earth, lay the first pieces of evidence that would provide clues to the origin and evolution of the Moon, and perhaps the Earth as well.

Using a pair of protective arms attached to the side of the container, a technician reached inside. With the unwieldy arms, he opened the box, removed a mesh covering and a strip of foil, and stepped back. Harvard geologist Clifford Frondell peered in. "Holy shit!" he exclaimed, as he stared at the dust-encrusted object inside. "It looks like a bunch of burnt potatoes!" Two days later, the first rocks had been cleaned, and scientists could see that they were basalt, formed from lava, similar to but lighter than terrestrial basalt.

At once, one of several controversies was solved. Some scientists held that the Moon had been torn from the infant Earth, others that Earth and Moon had formed together, yet others that the Moon had been captured by the Earth. Some had claimed that all the Moon's craters were made by impacts, others that they were produced by volcanoes. The first rocks showed that the Moon was enough like the Earth to suggest a common origin, and that molten rock had lain beneath its surface approximately 3.65 billion years ago.

IN QUARANTINE David Scott of Apollo 15 peers at the so-called "Genesis Rock" (above), which he and Jim Irwin had brought from the Moon. It turned out to be the oldest rock yet found.

ROCK HUNTER Harrison Schmitt, Apollo 17's geologist-astronaut, works close to the Lunar Rover during the last Apollo mission. He discovered some orange soil, which was found to contain ancient glass droplets.

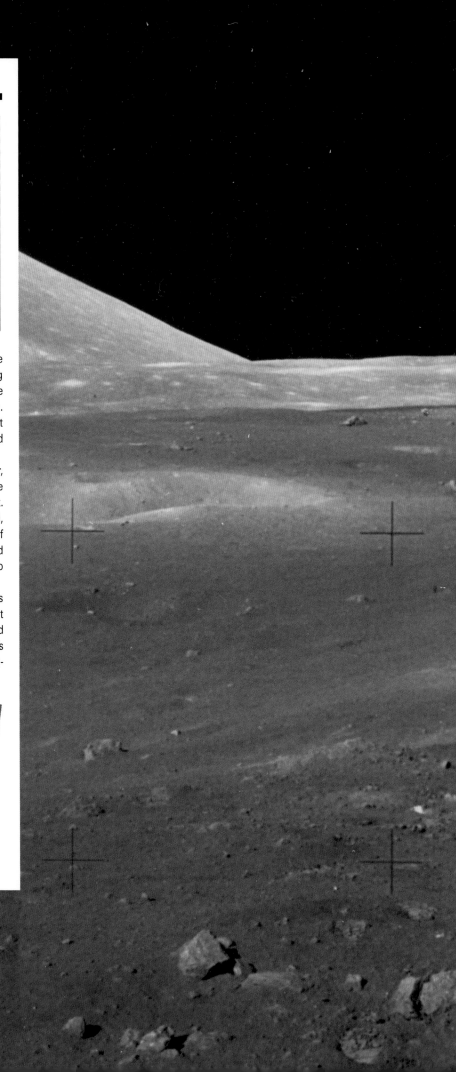

Office for six years, until surgery cured him, and he became the oldest man (at 47) to reach the Moon. His affliction turned out to have been a blessing in disguise. But for his ear problem, he might have been in the Apollo 1 capsule when it burned up; and he would have been on Apollo 13 if he hadn't been told to get more training after his years in the office. Things had moved on since he had belted himself into the Redstone almost a decade earlier: on Saturn V, the escape rocket alone was more powerful than the Mercury's Redstone had been.

For over 4 hours after their landing—the first on the lunar uplands—Shepard and crewman Edgar Mitchell trudged over the surface, hauling a rickshaw-like cart to gather rocks. While Stuart Roosa photographed the Moon from his lonely orbit, the men below collected 96 pounds of rocks.

Manned or unmanned?

So far, so good. But with the success of Soviet unmanned lunar probes, skeptics increasingly voiced concern at the price and risk of manned missions at all. The only possible answer was that men could do a better job than robots, which was what the last Apollo missions set about proving.

Apollo 15 astronauts David Scott and Jim Irwin brought a battery-powered Lunar Rover with them to explore the Apennine uplands. They landed near a mile-wide canyon known as the Hadley Rille, which they hoped might provide an open-cast slice

SOVIET MOON BUGGIES

Hard on the heels of Apollo 11, the Soviet Union sent a number of unmanned probes to the Moon. The success of these missions led many people to ask whether it was really necessary to continue sending men to the Moon.

In September 1970, Luna 16 soft-landed on the Sea of Fertility, dug up a $3^1/2$-ounce soil sample and sealed it into an ascent stage, which blasted off and returned the sample to Earth. In November, Luna 17 landed on the Sea of Rains. Down its ramp rolled a 1,666-pound device looking like an eight-wheeled casserole dish. Over the next ten months, this "Lunokhod" sent back 20,000 pictures and traveled

ROBOT SUCCESS The Soviet Union's unlovely, but practical, Luna 16 dug up some soil and returned it to Earth, thereby raising questions about the need for manned missions.

more than 6 miles. In 1973, a second Lunokhod would perform even better, while three years later a probe, Luna 24, would collect and return more lunar soil.

Such missions would cost NASA a fraction of the price of an Apollo mission, and carried none of the human risk. On the other hand, their results would be minuscule by comparison. A single Apollo mission collected hundreds of samples and placed dozens of long-lasting instruments. A dozen robot flights would be needed to match one Apollo mission. In the end, this logic overwhelmed the argument for robots. The Soviet lunar probe program, which had started as a face-saving venture, died with Apollo.

through time, like the Grand Canyon. With their rover, they traveled 17 miles and gathered 173 pounds of rocks, including the so-called "Genesis Rock," thought to date from soon after the Moon's creation.

Above them, Alfred Worden orbited with mapping cameras and spectrometers, which kept him busy. He also released a satellite that would remain in lunar orbit for another year, sending new data back to scientists on Earth.

Apollo 16 landed with its rover in the Descartes highlands, 18,000 feet above lunar "sea level." The ground crew, John Young and Charles Duke, set up an astronomical observatory and a cosmic-ray detector, and collected material from a crater $3/4$ mile across and 650 feet deep. Above them, in the command module, was Ken Mattingly, who

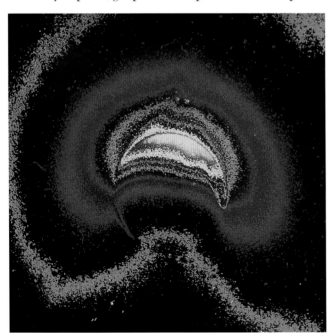

GASEOUS AURA The later Apollo missions looked at the Earth, the Moon and everything in between. This enhanced photograph from Apollo 16 reveals a tenuous corona of hydrogen around the Earth.

had helped save Apollo 13, and who never did develop German measles.

Apollo 17 landed 100 feet from its designated landing site, near a spot where landslides had brought material from the surrounding Taurus Mountains to the edge of the Sea of Serenity. This, the final Apollo mission, was commanded by Gene Cernan, with Ronald Evans as command module pilot. It also carried the first scientist to fly in Apollo, geologist Harrison Schmitt.

Schmitt had learned the new science of lunar geology under Eugene Shoemaker, who ran the U.S. Geological Survey's new astro-geology branch in Flagstaff, Arizona. Shoemaker—who later co-discovered the comet that crashed into Jupiter in 1994—had wanted to be the first scientist on the Moon. Instead, his protégé was chosen.

After setting up a batch of experiments, the two moonwalkers spent a whole Earth day on the surface, driving their rover 21 miles. Schmitt found some orange soil, its color deriving from minute glass droplets formed around 3.8 billion years ago. It was the final contribution to a rock collection that would take many years to assess, ensuring that for science, at least, the legacy of Apollo lived on.

LOOKING DOWN, LOOKING OUT

SCIENTISTS HAD ALWAYS BEEN INTRIGUED BY THE POTENTIAL OF SPACE EXPLORATION TO INSPIRE TECHNICAL INNOVATION AND EXPAND HUMAN KNOWLEDGE. TO MANY, THE SCIENTIFIC SPIN-OFFS FROM SPACE TRAVEL ARE THE REAL JUSTIFICATION FOR THE HUGE POLITICAL, MILITARY AND FINANCIAL COMMITMENT INVOLVED. WITH NEW TOOLS, SCIENTISTS HAVE REVEALED NEW WORLDS AND CHANGED OUR LIVES ON THIS ONE.

  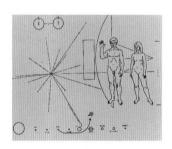

TOOLS IN ORBIT

BESIDES BEING USED IN SUPERPOWER RIVALRY, ROCKETS ALSO OPENED UP WHOLE NEW AREAS OF TECHNOLOGY AND RESEARCH

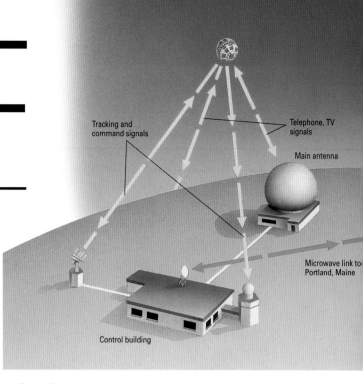

Manned space flight provided more than showbiz glamour. America had committed itself to winning both the arms race and the race to the Moon, and its successes in both areas spun off whole new industries. Over four decades, these spread the American-led revolution in information and electronics worldwide.

When communications satellites, known as comsats, first began to replace hilltop transmitters in the 1960s, they had some big disadvantages. First, they moved—and an orbiting transmitter traveling at 17,500 mph was difficult and expensive to track. Bell Laboratory's antenna, in the hills of Maine, which tracked the 1962 Telstar satellite, was 177 feet long and weighed 370 tons. Furthermore, satellites did not do their job to perfection: Telstar was in the right position for transatlantic calls for only nine months of the year.

Fortunately, a solution was at hand, which had been predicted almost 20 years before by the science-fiction writer, Arthur C. Clarke, the first person to foresee the true economic and social significance of the first tentative steps into space. In 1945, Clarke had pointed out that a satellite orbiting the spinning Earth at a certain distance above the Equator would always be located over the same place, appearing to be stationary in the sky—a phenomenon known as "geostationary orbit." Three satellites in geostationary orbit could receive signals from, and transmit signals to, the entire planet. It was a startling insight; while others dreamed of interplanetary travel, Clarke proposed that the future of astronautics lay in satellites that appeared to go absolutely nowhere.

KEEPING WATCH The tracking station for the Telstar satellite near Andover, Maine, had to work hard to stay in touch with a satellite that was not in geosynchronous orbit.

The path forward was revealed in 1959, when Explorer 6 briefly carried solar cells, vital for maintaining power in space, out to 26,000 miles. With a little nudge in the right direction, it could have become the first geosynchronous satellite. In 1963, Hughes Aircraft's engineering manager, Harold Rosen, persuaded NASA to launch such a satellite, Syncom 1, with a single two-way telephone channel. The satellite failed, as did its successor, but in 1964, Syncom 3 provided the first trans-Pacific link in time to show the 1964 Olympics live from Tokyo—for only 15 minutes a day to avoid undermining NBC, which had exclusive rights to televise the events, flying the footage back to the United States every day.

Almost immediately, 19 governments formed the International Telecommunications Satellite Organization (Intelsat), which put up the Early Bird satellite over the Atlantic in 1965. In May of that

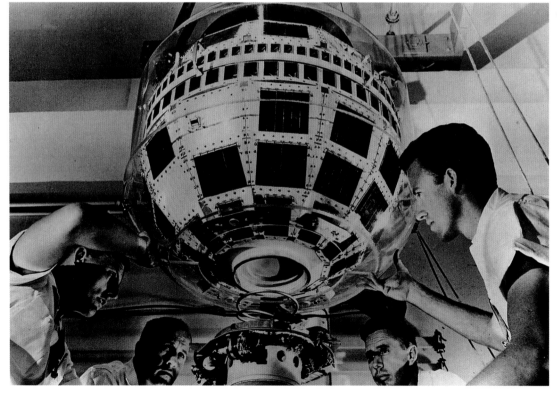

LINKING UP Technicians fit the Telstar comsat onto its rocket. It was launched in July 1962 to provide the first satellite link for telephone and television across the Atlantic.

1960

1962 Telstar comsat provides first transatlantic telephone and TV link

1964 Syncom 3 broadcasts the first live TV pictures across the Pacific

1965 Intelsat puts Early Bird comsat over the Atlantic; U.S.S.R. launches Molnya comsat

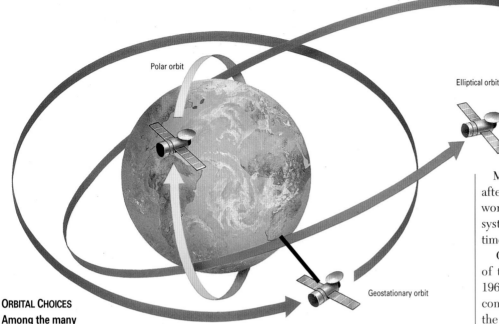

Polar orbit

Elliptical orbit

Geostationary orbit

ORBITAL CHOICES
Among the many possible orbits that a satellite can have, three are shown above. An elliptical orbit sends a satellite far from Earth, allowing it to cover a large area at any one time. A polar orbit stays close to the Earth and covers the whole planet in successive sweeps. A geostationary orbit holds the satellite locked above one point on the Earth's surface.

SOVIET LINK
Molnya ("Lightning"), launched by the U.S.S.R. in 1965, made telephone calls possible across the entire country. Its panels of solar cells are shown unfolded like the petals of an exotic flower.

year, TV coverage included political leaders in London, New York and Paris discussing the war in Vietnam. This was followed by "despun" satellites, which pointed toward one spot on Earth rather than broadcasting in all directions.

By 1969, Intelsat had global coverage. Sizes increased and prices tumbled. By the mid-1970s, Intelsat had 50 member states, with even bitter enemies cooperating for their mutual benefit: Israel and Egypt, India and Pakistan, even South Africa, which at the time was being spurned by most of the rest of the world because of its racial policies.

The Soviets, rejected by Intelsat at the beginning, went their own way. In any case, since their territory spanned northern latitudes only, satellites on equatorial orbits were of limited use. The Soviet Union's first telecommunications satellite, Molnya ("Lightning"), went up 17 days after Early Bird, creating the basis for the world's first domestic telecommunications system, vital for a nation that spanned 12 time zones.

Other large countries acquired satellites of their own—Australia, for example, in 1967, and Canada in 1972. By the 1980s, a completely new industry had developed in the United States, as satellite services were

continued on page 104

" CLARKE'S ORBIT

Arthur C. Clarke made his proposal for geosynchronous orbit in a little-read British journal, *Wireless World*, in October 1945:

"A true broadcast service, giving constant field strength at all times over the whole globe, would be invaluable...

"Many may consider the solution proposed in this discussion too far-fetched to be taken very seriously. Such an attitude is unreasonable, as everything envisaged here is a logical extension of developments in the last ten years...

"It will be possible in a few more years to build radio-controlled rockets which can be steered into orbits ... and left to broadcast scientific information back to Earth. A little later, manned rockets will be able to make similar flights with sufficient excess power to break the orbit and return to Earth...

"It will be observed that one orbit with a radius of 42, 000 km (26,000 miles) has a period of exactly 24 hours. A body in such an orbit, if its plane coincided with the Equator, would revolve with the Earth and would thus be stationary above the same spot on the planet...

"Using material ferried up from Earth, it would be possible to construct a space station in such an orbit... It could be provided with receiving and transmitting equipment ... and could act as a repeater to relay transmissions between any two points on the hemisphere beneath ... A single station could only provide coverage to half the globe, and for a world service three would be required, though more could readily be utilized. The stations would be arranged approximately equidistantly around the Earth."

1970 Bell Laboratory develops high resolution TV for a new generation of spy satellites

1972 The U.S. launches first of the Landsat series for investigating the Earth's resources

1975 European Space Agency (ESA) formed

1979 ESA launches first Ariane rocket

EYES ON ANTARCTICA

SATELLITES KEEP A CONSTANT CHECK ON THE WORLD'S HEALTH BY ANALYZING CHANGES IN THE SOUTHERN POLAR REGION

O ne odd consequence of the surge of information from space was that heavenly bodies, previously considered impossibly remote, became better known than some places on Earth. Even in the late 1990s, the far side of the Moon was better mapped than Antarctica. But with help from satellites, this frozen continent—as thoroughly concealed by its ice shield as Venus is by its clouds—began to reveal its secrets.

To some extent, Antarctica can be compared to another planet or the Moon in its inaccessibility and its hostility. Even in summer the South Pole averages –22° Fahrenheit. The cold and winds of winter, with six months of darkness, demand almost as much protection as is provided by capsules and space suits on a lunar journey. And since the Antarctic holds 90 percent of the world's ice, its influence on the world's climatic and ecological machinery is as all-embracing as the tides generated by the Moon and Sun.

Two areas, in particular, became a focus for the many nations that have research programs in Antarctica: the air above, and the ice beneath. Both provided evidence that human action was changing the Earth. In the early 1980s, the evidence they provided helped move the concept of "global warming" into the political domain, generating worldwide attention. This in turn inspired calls for more to be done, for changes in industrial practice, for further understanding and for international controls.

One area of research focused on ozone, which forms a diffuse layer in the upper atmosphere that screens out harmful ultraviolet rays. Ozone is destroyed by chemical compounds called chlorofluorocarbons (CFCs), widely used in refrigerators and aerosols. Starting in the 1970s, environmentalists fought to have CFCs banned from products. The environmentalists' position was greatly strengthened when, in 1985, satellite surveys

DANGEROUS OFFSPRING This detailed view of the north coast of Antarctica shows the leading edge of a glacier as it reaches the sea. Pushed ever outward by its own colossal weight, the glacier "calves" gigantic icebergs, which can take years to melt. Satellites allow researchers to maintain a constant census of the size and location of icebergs, for the sake of safety in world shipping.

INFRARED VISION Multiple scans taken with an infrared radiometer produced this image of the hidden landscape of Antarctica, including mountain ranges, glaciers and huge coastal ice shelves.

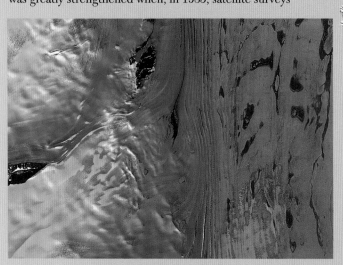

ICE FLOW Lambert Glacier, the largest glacier in the world, flows down to Antarctica's eastern coast, where it meets the Amery Ice Shelf. This infrared image shows the glacial ice as white and blue, while exposed rock appears black. On the right, successive shots of flow lines allow scientists to estimate the glacier's ice loss and speed of flow.

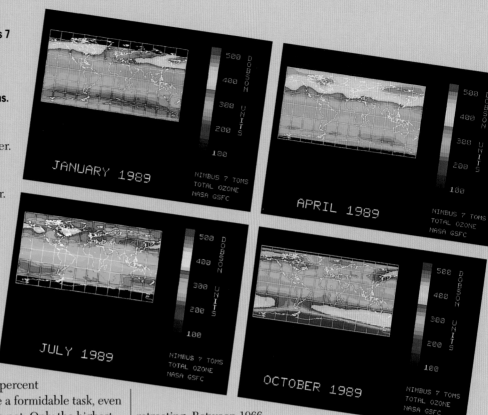

OZONE OVERVIEW Global surveying by the Nimbus 7 satellite revealed the world's changing patterns of ozone concentration, as this sequence from 1989 shows. Most worrying was the complete absence of ozone over Antarctica during its summer months.

revealed that the ozone above the Antarctic virtually disappeared each spring and summer. CFCs were doubly condemned, both as greenhouse gases that contributed to global warming and as destroyers of the ozone layer.

Though the immediate danger was small, the change was undeniably dramatic, for the "ozone hole" covered all of Antarctica, and more. If the hole was allowed to spread, human lives would be at risk. The international furor, and the persuasive effects of satellite imagery, resulted in a worldwide agreement, the Montreal Protocol of 1987, which deemed that CFCs would be banned by the year 2000.

The second focus for satellite-aided research was the ice cover. Antarctica is 30 percent larger than Europe, so mapping it would be a formidable task, even if its features were visible—which they are not. Only the highest mountains peep through the 1 mile-thick ice sheet. And that ice is dynamic. It is squeezed off the edge at speeds of up to 1/2 mile a year, contributing an estimated 2,000 billion tons to the oceans each year. This is equivalent to one-quarter of an inch added to global sea levels. Historically, this immense tonnage of water has found its way back onto the ice sheet, ensuring stability.

But if the Antarctic ice sheets should melt, they would raise sea levels by 230 feet, transforming the look of the world. Even a 1 foot rise would swamp vast tracts of populated land.

The chaotic nature of the Earth's climate makes it hard to measure global warming, or even confirm that it is a reality. One way this can be done, however, is to map and monitor Antarctica's ice. In 1972, the Landsat series of satellites made this possible. Aided by France's Systeme Probatoire d'Observation de la Terre (SPOT), launched in 1986, scientists could see features down to about 30 feet in size. Later, after the collapse of the Soviet empire, high-resolution photographs from Soyuz spacecraft have also become available. New satellites combine techniques—visual imagery, temperature sensors, radar scanning—to provide ever more detailed maps of a once-hidden world.

By measuring the size of the Antarctic ice shelves, scientists can see that the ice is retreating. Between 1966 and 1992, the Wordie Ice Shelf was seen to shrink from 750 square miles to 260 square miles. In 1996, one survey of nine shelves on the Antarctic Peninsula found declines in eight of them. Three had vanished completely.

Scientists are not eager to make sweeping conclusions. Perhaps such changes, which vary greatly from place to place, will reverse themselves. But the findings have sparked concern that much greater areas are at risk, like the whole West Antarctic Ice Sheet, holding 11 percent of the continent's ice.

Another task performed by satellites is to keep an eye on icebergs. In 1986, the Filchner Ice Shelf lost some 5,000 square miles all at once—an area the size of Northern Ireland that had taken 35 years to accumulate. The ice broke into three gigantic icebergs, one of which, called A24, drifted north. A24 was a block almost 60 miles across and 1,300 feet thick.

The iceberg drifted north, past the Falkland Islands, and finally broke up as it moved up the coast of South America. But it took three years to dissipate, and would have been a much greater danger to shipping if it had not been monitored by satellite.

PREFLIGHT CHECK Landsat 4 receives its final checks in July 1982. Although the satellite's main scanner—which could resolve detail down to 100 feet—failed after 11 years, its secondary scanner, with a resolution of 260 feet, remained operational. This satellite has been vital in monitoring Antarctic ice movements.

combined with cable communications systems. HBO offered movies, while CNN provided around-the-clock news.

Comsats offered revolutionary benefits to the world's underdeveloped nations as well. Vast and impoverished countries, which could not contemplate the task of laying

cable networks, could leap into the world of instant communications. Soon Indonesia, Brazil, India and the Arab world were all embraced by the new "global village."

Spying on the enemy

Since the early days of satellites, space-based military reconnaissance has made huge advances in accuracy. Through the 1960s, the CIA ran a "bucket-drop" program,

RAIN FOREST LOSS This false-color Landsat view of the Brazilian Amazon contrasts virgin forest (dark green) with cleared land (pale green and pink). Year by year, satellites have shown the dark green areas shrinking.

retrieving film canisters ejected from orbit into the atmosphere. Space cameras, designated as KH (Key-Hole), became increasingly sophisticated. Low-flying satellites could take photographs showing objects a mere 18 inches in size. In 1970, Bell Laboratory invented the charge-coupled device—in effect, television coverage from space—capable of showing details as small as 6 inches. From 500 miles out, American analysts could see dockside cables in a Soviet Black Sea shipyard.

With such "spies-in-the-sky," the Soviet Union could hide little. The previously secret site of Plesetsk, near Archangel, was identified as a major source of missile launches, and Severodinsk as the main construction site for ballistic-missile submarines. American analysts saw antiballistic missile sites guarding Moscow and Leningrad. These findings alone were enough for American leaders to justify their expenditure on orbiting space technology. It saved money, as well as unnecessary fear. In the words of one analyst, Jeffrey Richelson, satellite reconnaissance was "one of the

most significant military developments this century, and perhaps in all history . . . The photo-reconnaissance satellite, by damping fears of what weapons the other superpower had available and whether military action was imminent, has played an enormous role in stabilizing the superpower relationship."

Sunshine, rain or snow

Weather satellites ushered in yet another revolution. In 1960, TIROS 1 had revealed for the first time the global nature of cloud cover and related weather fronts. From orbit, the Earth was a giant weather map, and details were upgraded with every new version of TIROS. In January 1965, TIROS 9 was placed in a polar orbit and recorded the whole Earth, except Antarctica, in 480 shots. Soon, upgraded satellites could transmit their views automatically, on demand, via a receiving station and fax.

Through the 1970s, some 400 weather satellites were placed in orbit, servicing 40 countries daily—and nightly, thanks to infrared imaging. The first geosynchronous weather satellites provided wide-angle views: one set of shots showed six typhoons and hurricanes developing and moving all at once. For the first time, meteorologists

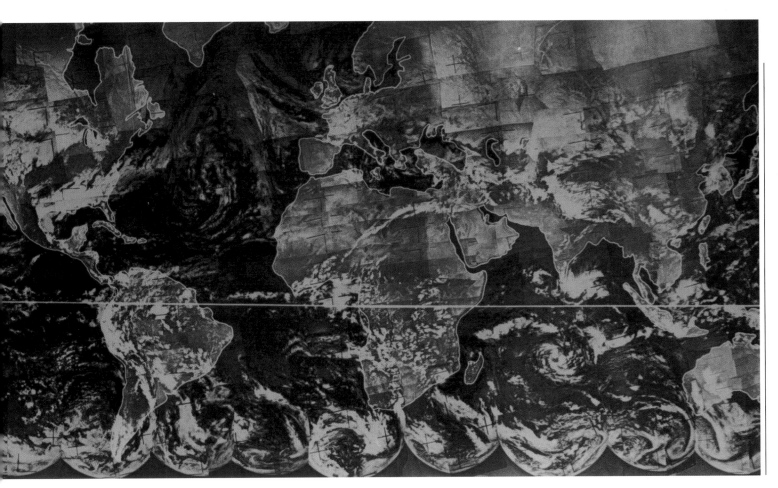

WEATHER WATCH TIROS 9 required 12 orbits to compile this, the first complete image of the world's weather (above). The semicircular "bubbles" at the bottom of the image mark the edge of each sweep. Since then, satellites have monitored storms such as Hurricane Allen (right) crossing Texas in August 1980.

around the world could monitor these destructive storms from birth to death.

Meteorologists could keep an eye on other elemental systems as well—snow and ice cover, currents, vegetation, fault lines— opening new chapters in half a dozen Earth sciences. They could see a glacier-surge in Alaska, an ice shelf breaking up in Antarctica, Saharan sands spreading southward, newly cleared lands in the vanishing Amazon rain forest. From these surveys came the first suggestion that the planet was undergoing fundamental change caused by human activities and industrial pollution—the first inklings of the existence of global warming.

Europe enters the ring

These developments showed other nations that American know-how could dominate this new area of commercial expansion forever, if it remained unchallenged. Europe, which had

NEW CHALLENGES The launch of the French Ariane rocket in 1979 (above) announced the arrival of the European Space Agency as a major player in space. Half the world's launches still came from once-secret Plesetsk (right), 400 miles north of Moscow, but the balance of power was shifting.

been little tempted by manned missions, saw the benefits of commercial space flights and opted for a space race of a very different kind. In 1961-3, ten European nations founded two organizations to create both a launcher—Europa—and projects to put in it.

But Europe had no overall command structure like NASA. Nor could it provide massive suppliers such as Douglas Aircraft, so the only thing that went into orbit was the budget. In 1973, after costing 3½ times its original estimate, Europa was canceled.

By then, France already had its own space program. Under the ambitious guidance of its president, General Charles de Gaulle, France had earlier developed its own strategic strike force independent of NATO, which was dominated by the United States. De Gaulle wanted nuclear weapons, nuclear submarines and rockets. As the United States and the Soviet Union had already proved, venturing into space was a logical extension of these ambitions. When the tiny

Diamant satellite launcher put France's first satellite into orbit in November 1965, it showed the world that France could succeed where Europe could not.

The French also had one huge advantage: they ruled over French Guiana, a perfect launch site lying almost on the Equator. In 1975, with France taking the lead, 14 European nations combined to form a new organization, the European Space Agency (ESA), which backed France's own launcher, Ariane. In successive versions, Ariane evolved into a flexible system that used a combination of liquid and solid boosters. With Ariane's first flight in 1979, Europe was finally ready to compete with America.

REVEALING THE INNER PLANETS

THE RACE FOR ASTRONOMICAL KNOWLEDGE BROUGHT MERCURY, VENUS AND MARS INTO THE REALMS OF PRACTICAL SCIENCE

HARD LANDING Looking like the archetypal bug-eyed monster, the Soviet Union's Venera 3 could have been built to appeal to the denizens of another world. It became the first man-made object to reach the surface of another planet when it crashed on Venus in March 1966.

Of all the planets, Venus, the "Evening Star" and the "Morning Star," has always been the most alluring and the most mysterious. It is the nearest planet to Earth, the most similar to Earth in size, and the brightest. It is also the most obscured of the inner planets, for its brightness is caused by sunlight reflecting off clouds that permanently surround it. Until the dawn of the space age, no one had any idea what the clouds concealed, and there was no way of telling how long the planet took to rotate.

Since Venus lies closer to the Sun than Earth does, many guessed it was tropical, perhaps covered in forests similar to those on Earth some 225 million years ago, with giant ferns, dragonflies and swamp-dwelling amphibians. On the other hand, perhaps the clouds reflected all the heat, leaving it icy. No one knew whether a probe would land in a desert, an ocean, a dense, giant forest or an ice field. The astronomer Fred Hoyle even

SAILING BY The Mariner 2 was the first probe to reach Venus. It flew past in 1962, revealing the planet's hellish atmosphere and its odd reverse spin.

suggested Venus could be covered by oceans of oil. The Soviets decided to be the first to find out.

Venus's deceptive beauty

The first Soviet probe, launched in 1961, failed, as did America's first mission. In late 1962, however, the United States's Mariner 2 flew past Venus, recording an atmosphere of carbon dioxide with clouds rich in sulphuric acid and a surface temperature of 750° Fahrenheit, hot enough to melt lead. The temperature was almost as high on the night-side of the planet, showing that the atmosphere acted like a quilt.

Venus proved odd in other respects too. It spins very slowly, taking 243 Earth days to complete one rotation; and it spins backward, or away from its orbital path around the Sun, so that its sunrise is in the west. Because it orbits the Sun in about 225 days, and because of its retrograde motion, the Venus day is 117 Earth days long. Yet its upper atmosphere forms a jet stream that circles the planet every four Earth days.

These early findings sparked new insights into Earth's history. Venus was roughly the same size as the Earth, and presumably had similar origins. What had happened to make one a womb for life, the other a hell? Apparently, something had prevented Venus from cooling, ensuring that all the carbon dioxide remained aloft to create a runaway greenhouse effect. In the heat, the hydrogen that would otherwise have produced water boiled away into space, leaving an arid oven of a planet. The Earth had been saved from a similar fate by a stratosphere that trapped water molecules. Thereafter, emerging life produced an oxygen-rich atmosphere.

Astronomers were eager to know what Venus's surface was like: whether there was

anything solid down there, anything liquid, whether violent winds shrieked, whether the Sun penetrated or whether stygian darkness ruled. Some astronomers predicted that, if anything could be seen, the dense atmosphere would distort the view, so that things would appear curved, as if in a goldfish bowl.

Again, the Russians took the lead, with some failures, and some dramatic successes. Between 1961 and 1966, nine Venus probes failed, either crashing after takeoff or failing en route. In 1966, Venera 3 crashed onto Venus, becoming the first man-made object to reach the surface of another planet. But in October 1967, Venera 4 parachuted into the planet's upper atmosphere, transmitting for an hour and a half before being snuffed out by air pressure, which it measured as 90 times that on Earth. Any human visitor to Venus would be poisoned, broiled, eaten by acid and squashed simultaneously. Later probes were made considerably tougher. In

THE VEIL STRIPPED In 1990, the Magellan probe sent back a series of high-resolution radar images, which showed the Venusian landscape in breathtaking detail.

1970, Venera 7 landed and survived for 23 minutes, transmitting the first messages from the surface of another planet. In 1972, Venera 8 lasted a few minutes longer.

Then, in 1975, Veneras 9 and 10 were sent up, carrying refrigeration units and cameras. They endured for an hour before the heat overwhelmed them, but this was long enough to send back some black-and-white pictures. These images revealed a twilight "stony desert" of dust and rocks averaging about 3 feet across. To everyone's surprise, the jet streams above had little effect below. With such enormous air pressure, even a breeze would be as devastating as terrestrial sea waves blown by a gale, yet the rocky ground showed little evidence of abrasion. In 1982, Veneras 13 and 14 sent back color pictures showing rocks glowing orange in a fiery fog.

The next step was to map Venus. In this pea-soup world, the only way to do so was with radar. Several probes in the 1980s started this task, most notably Russia's two Vega projects, which included experiments devised by a half-dozen other nations. But the main details had to wait until America's

FRIED PROBE The early Venus probes succumbed quickly to the planet's intense heat. This shot from Venera 13 shows rocks, pebbles and soil under an orange light—Venus's atmosphere strips away blues and yellows.

SCIENCE DREAMING

From the early days of space exploration, scientists, writers and artists have inspired each other with ideas about what the planets might actually be like. Until close-up photography killed their dreams, artists creating illustrations for popular science-fiction magazines were free to improvise within the current limits of knowledge.

Well into the 1970s, for instance, astronomers thought that Mercury kept one face permanently turned toward the Sun, creating three zones—a night side that was among the coldest places in the Solar System, a day side hot enough to melt metal into streams, and a twilight zone. For writers of science fiction, this third, twilight zone was very likely to be inhabitable.

Venus, with its face permanently obscured by clouds, offered precious few facts to restrict the imagination until the first Venera probes landed in the 1960s. Until then, it was commonly assumed, in line with the Swedish astronomer Svante Arrhenius, that Venus was still a "young planet," a wilderness covered with luxurious, semitropical vegetation, as the Earth was during the Carboniferous period, about 225 million years ago.

Mars, on the other hand, was thought of as an ancient world of deserts and badlands, the very image of decay. Countless writers and artists portrayed the "canals" that the astronomer Percy Lowell claimed to have seen. Dozens of sci-fi magazine covers showed struggling Martians trying desperately to revive their world by guiding polar meltwaters across its desiccated surface.

The huge distances of the outer planets from Earth allowed even freer rein to the imagination. Artists dreamt up fantastic animals and mythical landscapes on the surface of Jupiter, for example. In these settings, human beings could play out time-honored epics—ancient struggles with a veneer of scientific respectability.

As the space age provided ever more facts, however, the great days of the sci-fi magazines ended, and space began to appeal to artists who were inspired by what was really out there.

Magellan probe, launched by the Space Shuttle, arrived in 1990. Using a 12-foot dish, Magellan bounced radar pulses off the surface. The resulting images revealed a strange world. Over half the planet is one huge plain. The rest consists of uplands. In the north, a lava plateau rolls up to the volcanic cone, as high as Mt. Everest, that dominates the Maxwell range. The south consists of another great plateau. Craters are numerous, lava flows common.

Venus remains significant for science as a world that might have been. But no one expects that any astronaut will ever need to risk its appalling conditions.

The ancient waterways of Mars

Of all Earth's neighbors, Mars offered the strongest possibility of being Earth-like. Its surface could be seen directly from Earth. Its reddish tinge showed variations, and its white, icy poles shrank and grew, suggesting weather and seasons. It was a close neighbor, only half the size of the Earth, but with a spin about the same as ours. There was enough to suggest some intriguing similarities—and to inspire

one of the most peculiar and long-lived controversies in the history of astronomy.

The controversy concerned the "canals" of Mars. The idea was first popularized in the 1880s by the Italian astronomer Giovanni Schiaparelli, after he mapped what seemed to be strange lines, some of them thousands of miles long, crisscrossing the surface of the planet. He called them *canali* and suggested they were geological features formed by water flowing from the icy poles.

The idea inspired a wealthy American, Percival Lowell, to establish his own observatory in Flagstaff, Arizona, and to dedicate most of his life to the subject. As if obsessed by an anglicized version of the Italian word, Lowell became convinced that the *canali* were actual canals—artificial waterways constructed by Martians. He guessed correctly

that Mars was a desert planet, but went on to suggest that the Martians were struggling to irrigate their world. He explained that the lines were canals and the vegetation that grew along them. He concluded: "A mind of no mean order would seem to have presided over the system we see."

In reality, Lowell's ideas were based on optical illusions and wishful thinking. Neither he nor Schiaparelli could possibly

have seen such fine detail with an Earth-based telescope. Nevertheless, the idea was so appealing that it was in vogue for many years, inspiring countless science-fiction stories and an enduring hope that life would one day be found on Mars.

In the 1960s, the coming of the space age added several twists to the saga. With Mars, Soviet efforts proved remarkably vulnerable. Two Soviet craft landed, but survived for only 20 seconds. Mars 5, an orbiter, sent back some pictures in 1973, but gave up after 20 orbits.

American probes were more successful. In 1965, on a fly-by mission, Mariner 4 sent back 21 pictures and a mass of additional information. By measuring variations in Mariner's signals as it vanished behind Mars, scientists discovered that Mars's atmosphere was far more rarefied

INTERIOR VIEW Mars's low density, weak magnetic field and small size suggest that it has the following make-up: a small magnetic core and a thick mantle, beneath a crust twice as deep as the Earth's.

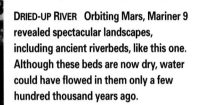

DRIED-UP RIVER Orbiting Mars, Mariner 9 revealed spectacular landscapes, including ancient riverbeds, like this one. Although these beds are now dry, water could have flowed in them only a few hundred thousand years ago.

than they had predicted—$1/200$ of that of the Earth. There was no sign of canals, and there were a great many craters. It seemed likely that the polar caps were not made of water-ice, but of frozen carbon dioxide. When Mariners 6 and 7 flew past in 1969, a few days after the first Moon landing, they backed up previous findings. "A dull landscape, as dead as a dodo," commented one member of the Mariner team.

But he was wrong. By pure bad luck, those two Mariners had recorded the most

1982 Veneras 13 and 14
(U.S.S.R.) send back first
color pictures of Venus

1990 Magellan
probe (U.S.) maps
Venus by radar

1997 Pathfinder and
Sojourner rover (U.S.)
on Mars

boring bits of the planet. When Mariner 9 dropped into Mars orbit in 1971, becoming the first man-made object to circle another planet, its images sparked a revolution in the study of Mars. For one thing, Mars was by no means dead geologically—the probe's cameras revealed a dust storm that blanketed the whole planet. Indeed, a Russian lander that descended into the storm transmitted for only 20 seconds before being wiped out by wind and dust.

When the dust cleared, the sights were staggering: a volcano 375 miles across and three times the height of Mt. Everest; craters galore; a great rift valley four times deeper than the Grand Canyon, slashing across half a hemisphere; a lava flow 400 miles long; and, most startling of all, river valleys. They were all dry, for no liquid water can exist on Mars today, but they have led scientists to think that life might have survived on a small scale, as microbes survive in deserts and sulphur springs on Earth.

The excitement continued with the twin Viking landers that arrived in 1976. Each was equipped with cameras and an on-board biological laboratory to test for life in the

VANISHING CRATERS Whereas impact craters on the Moon can last for eons, those on Mars are in a state of flux. The planet's violent atmosphere erodes them and fills them with dust, making them slowly disappear.

soil. Some scientists even hoped that the cameras would see Martian bugs, which might draw moisture from the Martian frosts and protect themselves with silica parasols. One probe parachuted down into a basin where water had once flowed; the second Viking came down 4,600 miles away in an area noted for its winter frosts.

The two Vikings recorded red, rocky landscapes, light winds and a maximum temperature of −24° Fahrenheit. In all, they confirmed Mars's grimness. If all the water on Mars was gathered together in liquid form, it would make just one large lake. There is no rain. In the thin air, the temperature seldom rises above freezing. The top mile of the planet's crust is probably permafrost. The biological experiments found no signs of life, even

though they were capable of detecting organic remains at concentrations 1,000 times less than the Earth's most barren deserts. "That's the ball game," commented Viking's chief scientist glumly. "No organics on Mars, no life on Mars."

Even so, the Vikings provided enough evidence to suggest a possible biography for Mars. It formed in the same way as Earth, with a crust, a mantle and a core. Asteroid impacts blasted the surface, churning it up to a depth of 1 mile or more, releasing enough oxygen to make water, a process aided by volcanoes. Mars could well have had a dense atmosphere, and millennia of rain. But its gravity could not hold its atmosphere in place, so the gases escaped into space. Water also evaporated into space, except perhaps for some remaining in underground pockets. Eventually, the water was exhausted, perhaps relatively recently, only hundreds of thousands of years ago; if all the water had disappeared earlier, the water courses would have been eroded away by the planet's fierce dust storms.

Interestingly, although the Vikings found no life on Mars, they did discover that the planet's soil might be capable of supporting life. In 1996, scientists found evidence of what could be microscopic microbial fossils in pieces of Martian rock that arrived on

MARTIAN SUNSET This twilight view from Viking 1 shows a stark Martian landscape of rocks and dust. The lander found no life, but could explore only a tiny area.

PATHFINDING ON MARS

A ROBOT ROVER WRITES AN AGENDA FOR ONE FUTURE STRAND OF PLANETARY RESEARCH—AUTOMATED, MINIATURIZED, SOPHISTICATED, CHEAP

O n July 4, 1997, the United States landed a spacecraft on Mars containing the first Mars rover. The lander, known as Pathfinder, came down from orbit onto a rock-littered plain, Ares Vallis. It immediately began sending back 1,700 pictures of its surroundings, forming a wrap-around panorama of startling immediacy. Pathfinder even had its own Web site on the Internet. Users could see pictures captioned "Live from Mars."

Pathfinder lowered several ramps, down one of which crawled the 24-pound six-wheeled rover, Sojourner. It inched its way down the ramp, and used a spectrometer to "sniff" at the dust under its wheels. It reported that the soil was iron-rich, just like that found by the Viking landers 21 years earlier. Apparently, the rusty Martian soil, punctuated by sand dunes, is scattered planet-wide by its winds.

Sojourner was next directed to examine a 10-inch high rock a couple of feet away. Again, the rover deployed its spectrometer, analyzing the rock scientists had already dubbed Barnacle Bill. In the words of one jubilant scientist: "Sojourner sort of nestled up and kissed Barnacle Bill." To the scientists' surprise, the rock seemed to be like an andesite, a type of volcanic rock named after the Andes mountains. On Earth, such rocks are formed by being repeatedly melted and solidified as part of the process that forms continents. Yet Mars was thought to lack such activity.

In a week, Pathfinder and Sojourner returned more than 16,000 images and made 15 chemical analyses. The rocks and pebbles at the landing site suggested that the landscape had been created by running water. "Water, of course, is the very ingredient that is necessary to support life," said one of the leading project scientists,

INTERPLANETARY KISS Wearing a cap of solar cells, Sojourner nudges up to a rock nicknamed Yogi Bear because its shape resembled a reclining bear.

Dr. Matthew Golombek. "And that leads to the $64,000 question: Are we alone in the universe? Did life ever develop on Mars? If so, what happened to it? If not, why not?"

After six weeks, the lander's battery failed and contact ended. But even if Pathfinder and Sojourner are dead, their findings established the viability of low-cost planetary missions, providing evidence that is vital for a better understanding of the evolution of the Solar System, the Earth, and life itself.

READY FOR ACTION Pathfinder stands on the Martian surface with its ramps lowered and its camera surveying the surrounding landscape.

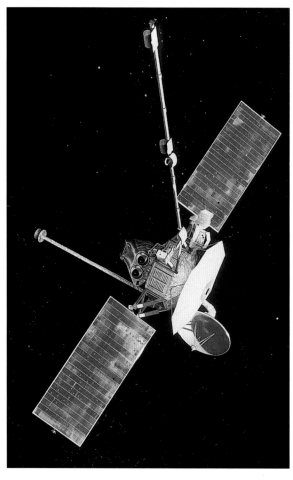

Earth as meteorites. (They had been blasted into space by other meteors striking Mars in the first place.) The findings remain controversial, but some scientists still think it possible that scattered oases could harbor minute and hardy life forms on Mars—living fossils left over from a fertile and well-watered past.

The following year, 1997, saw the most ambitious Mars landing yet with the Pathfinder mission; but again, no living creatures were found.

Mercury the mysterious

The planet nearest the Sun, Mercury, has been an astronomical enigma for centuries. It is bright, fast-moving and almost impossible to examine against the Sun's blinding light. Suddenly, in 1974, its veil of mystery was raised.

America's Mercury mission—the only one to date—was actually a by-product of a longer-term agenda: an unmanned mission to the outer planets. To get that far would require a technique that employed the gravitational attraction of one planet to grab a passing probe and accelerate it on its way in a slingshot effect. The strategy demanded extreme accuracy. Mariner 10, which used Venus's gravity to increase its own speed and fling it on its way to Mercury, was a trial run for the later deep-space Pioneers and Voyagers.

Mariner 10 was a success. It whipped past Venus and into solar orbit so accurately that it flew past Mercury three times, making a

detailed photographic record of much of the surface before its transmitter died.

Mariner 10 revealed how odd Mercury is. The planet rotates three times while going around the Sun twice, giving it a day lasting two Earth years. It also has a rather eccentric orbit. Perhaps these oddities were caused by the same ancient event—an encounter with a particularly large asteroid—that left a scar, called Caloris Basin, 810 miles across.

The impact set up such a shock wave that it jumbled the surface on the other side of the planet, immediately opposite the impact point. The collision might also have been enough to bump Mercury into its eccentric orbit, and to melt so much rock that it acquired tides. Perhaps the friction of these rocky tides, slopping back and forth as Mercury cooled again, slowed its spin until it fell into its unique 3:2 harmonic relationship with its orbit time.

Mercury seems to have formed with a hot core, in the same way as the Earth, for it has a magnetic field, probably produced by a liquid-iron core spinning as the planet spins. Yet it looks more like the Moon, pockmarked with ancient craters. Like the Moon, it has no atmosphere to speak of. Nor does it show any signs of continuing volcanic activity.

Yet it is also different from the Moon in that it has fewer huge craters and no mountains; it has more smooth plains, and masses of escarpments. Apparently, when the Solar System was forming, the bigger meteorites and asteroids were drawn away from Mercury by the overwhelming force of the Sun's gravity. Unlike the Moon, Mercury seems to have shrunk as it cooled, thereby acquiring its many wrinkled escarpments.

DEAD WORLD The Mariner 10 (above), with its 25-foot solar panels and telescopic cameras, swung by Mercury three times before its transmitter stopped working. In a mosaic of images sent back by Mariner 10 (below), Mercury appears Moon-like, with its thousands of impact craters. However, the planet has only one "sea." It also has fewer large craters and no mountains. Instead, it has plains and escarpments that are hundreds of miles long.

THE GASEOUS GIANTS

A STROKE OF INGENUITY DEMONSTRATED THAT THE OUTER PLANETS WERE WITHIN REACH—WITH DRAMATIC RESULTS

NASA's most impressive interplanetary missions were made possible by the work of Gary Flandro, an analyst working at the Jet Propulsion Laboratory (JPL) in the mid-1960s. For an ambitious organization such as JPL, one obvious avenue of development was to look outward to Jupiter and its more distant companions, Saturn, Uranus and Neptune, the giants of the outer Solar System. However, a huge, 600 million-mile gap—six times the distance from the Earth to the Sun—opens up between Mars and the next planet, Jupiter. At the speeds reached by NASA's most powerful launch vehicle, the Titan III, a probe would take ten years to reach Jupiter, 30 to reach Neptune. It seemed a hopeless task.

Flandro knew, however, that when Mariner 4 had flown past Mars in 1965, it picked up an extra 2,000 mph, accelerated by Mars's gravity. He reasoned that if Jupiter was approached from the right direction, its gravity would increase a probe's speed by much more than Mars's had. The acceleration would be the equivalent of another rocket stage, adding enough speed to whip a probe onward to the next planet, Saturn. Moreover, Flandro saw that the outer planets would, some ten years later, enter a rare alignment. With the right course, probes could use each planet in turn to gain enough speed to reach the next, spiraling outward ever faster.

If these missions were to succeed, however, JPL needed some practice runs in order to build its experience in guidance techniques. One such practice mission was Mariner 10, which used Venus's gravity to reach Mercury in 1974. Two others were Pioneers 10 and 11, which visited Jupiter in 1973 and 1974 respectively. The Pioneers showed that the journey to Jupiter could be cut to about a year and a half. In addition, they showed that a spacecraft could survive a journey through the asteroid belt, a collection of millions of rocks orbiting between Mars and

GRAVITY'S SLINGSHOT This diagram shows how Pioneers 10 (green line) and 11 (yellow line) hurtled into Venus's gravity field and out again, picking up enough speed to reach Jupiter in 1^1/2 years.

Jupiter. Until then, no one was sure if this was remotely possible.

Pioneer 10 would have a special status. Besides being the first probe to look at Jupiter close up, it would go on to become the first man-made object to leave the Solar System. Pioneer 11 would follow in its wake. These two wanderers would lead the way into the immensities of interplanetary space. Unless destroyed by impact with an asteroid, they would travel for eons, perhaps eventually approaching another star system.

MESSAGE FROM EARTHLINGS Pioneers 10 and 11 both carried this plaque designed by Carl Sagan. It shows a hydrogen atom (top left); this is the key for decoding its message, which details the probes' origins in time and space.

Perhaps some sophisticated civilization would notice them, and retrieve them.

Inspired by this possibility, the astronomer Carl Sagan designed a plaque that was attached to each craft. The design looked like a weird array of symbols and Morse code. In fact, it was devised so that any intelligence capable of space travel could interpret it. Using the hydrogen atom, the universe's fundamental building block, to provide a measure of distance and time, it pinpointed the Sun with respect to other stars, and the Earth among the Sun's planets. It even showed a naked man and woman, the man with his hand raised in greeting.

Into the unknown

So it was that in March 1972, Pioneer 10 headed on its way toward Jupiter, to open a new chapter in exploration. It was followed

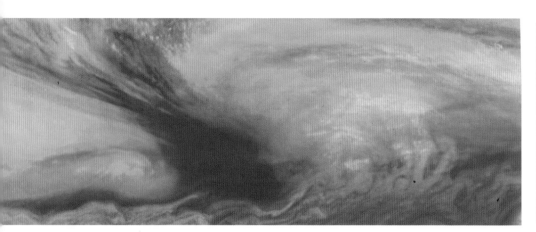

CHEMICAL SOUP Pioneers 10 and 11 revealed that Jupiter is covered by violently swirling clouds of gas (left), which are spun into bands by the planet's rotation. One feature is semipermanent—the vortex known as the Great Red Spot, seen below with two of Jupiter's moons, Io and Europa.

just over a year later, in April 1973, by Pioneer 11 on its way to both Jupiter and Saturn. If these were successful, two more complex devices, Voyagers 1 and 2, would follow later in the decade. Together, the four

1972 Pioneer 10
launched

1973 Pioneer 11 launched;
Pioneer 10 reaches Jupiter

1974 Pioneer 11
reaches Jupiter

1977 Voyagers 1
and 2 launched

1979 Voyagers 1 and 2
reach Jupiter; Pioneer
11 reaches Saturn

IN ORBIT AROUND JUPITER

The Galileo mission to Jupiter was one of the most jinxed, and yet most successful, of missions. It took 12 years for its six-year journey to begin.

Jet Propulsion Laboratory won approval for this orbiting probe in 1977. Originally it was to have been blasted on its way by a Titan III. But the Space Shuttle became its chosen carrier. Four months before Galileo's scheduled launch in May 1986, the shuttle *Challenger* blew apart just after takeoff, setting back NASA's schedules for the next two years. Galileo was transported back to California to await a new launch vehicle. It was finally sent up on its long journey in October 1989.

As Galileo began a long swing past Venus and back around the Earth to pick up speed, scientists discovered that during the delays the ribs of its umbrella-like antenna had bonded together and refused to unfold. With only a secondary antenna working, the speed of its transmissions were 3,000 times slower than planned. But the delay on Earth had seen great advances in electronics. Galileo's computer could compress its data tenfold, and also edit out of its pictures the blackness of space.

When Galileo reached Jupiter in December 1995, it turned in a magnificent performance. As it approached, it divided in two. One section parachuted down through the atmosphere, collecting

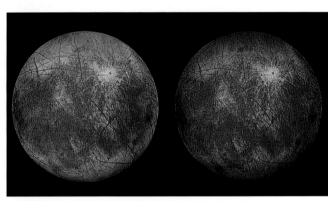

SURFACE ICE Europa's surface looks like a cracked eggshell. It is apparently fractured ice with no surface features.

data for almost an hour before it was buffeted and crushed into silence. The other section went into orbit on a two-year mission. It then began a new series of observations focusing on Europa, Jupiter's sixth moon, whose icy surface may conceal liquid oceans. Pictures taken in 1997 showed mottled terrain and the scars of a fault line, all suggesting recent geological activity. Over the next two years, Galileo would make eight passes of Europa, then several more of Callisto and Io.

would throw light on some of the secrets of Earth's distant, giant siblings.

Jupiter has long fascinated astronomers. The largest body in the Solar System, it has 300 times the mass of the Earth and $2^{1}/2$ times more mass than all the other planets put together. It was known to be a great, soft ball composed mainly of hydrogen and helium, with a density one-quarter that of the Earth. Its atmosphere was apparently torn by violent winds, probably induced by its fast spin—once every 9 hours 50 minutes. Vast air currents circle the planet at different speeds, creating a series of bands, as well as a hurricane-like vortex, known as the Great Red Spot, which itself is larger than the Earth.

Of equal interest were Jupiter's moons. Four of them are large enough to be seen through a small telescope, and were first observed by Galileo in the winter of 1609-10. Two of them (Ganymede and Callisto) are larger than the planet Mercury, while the other two (Io and Europa) are about the size of our Moon. A small fifth moon was discovered in 1892.

Saturn, Uranus and Neptune, all smaller versions of Jupiter, were known to have hydrogen-rich atmospheres, with other gases that colored them—Saturn is yellow and orange, Uranus green, and Neptune blue. All were known to have several moons. Uranus was odd in that it was tilted over, so that it seemed to roll along its orbit, with first one pole then the other exposed to the Sun. Beyond this, very little was known about them, while even farther out lay tiny Pluto, an enigmatic dot barely larger than our Moon, that would not be in line for a visit.

The really startling feature of these outer planets was Saturn's system of rings. These beautiful, regular, multicolored disks, with an overall diameter of about 170,000 miles, and easily visible through an Earth-based telescope, had been a mystery until the mid-19th century. Astronomers assumed that they were either liquid or solid until, in 1848, a French mathematician, Edouard Roche, showed that this was not possible; if the rings were liquid or solid, they would have been disrupted by Jupiter's gravity.

Scientists concluded, correctly, that the rings were probably composed of particles of some sort.

Pioneer 10 succeeded magnificently. It also showed that crossing the asteroid belt was not as hazardous as had been feared.

COMPLEX HALO Saturn's colorful rings, as revealed by Pioneer 11 in 1979, turned out to be more varied than expected. Later, the Voyager probes revealed seven distinct bands, each subdivided into several subsets.

The closest it came to a major boulder was 5.5 million miles. Once the Pioneers had shown that the gravitational "slingshot" maneuver could work, JPL and NASA sent the Voyagers on their $1^{1}/2$-year journeys in 1977. They both reached Jupiter in 1979. Both journeyed on to Saturn, reaching it in 1980-1. Voyager 1 then headed out of the Solar System, while Voyager 2 swung on to Uranus (1986) and Neptune (1989).

New moons, new rings

In combination, Pioneers 10 and 11, and Voyagers 1 and 2, provided astonishing new information for scientists to pore over. Even the first, Pioneer 10, produced wonderfully detailed shots of Jupiter's swirling Red Spot and majestic thunderstorms. Jupiter was shown to have a hot center, some $50,000°$ Fahrenheit, hidden by an ocean of liquid hydrogen. It had a strong and vast magnetic field that reached out as far as Saturn.

Yet the most intriguing pictures were those of Jupiter's moons and rings. The probes saw four new small moons—which brought the total to 16—and two previously unknown rings. They also revealed that the

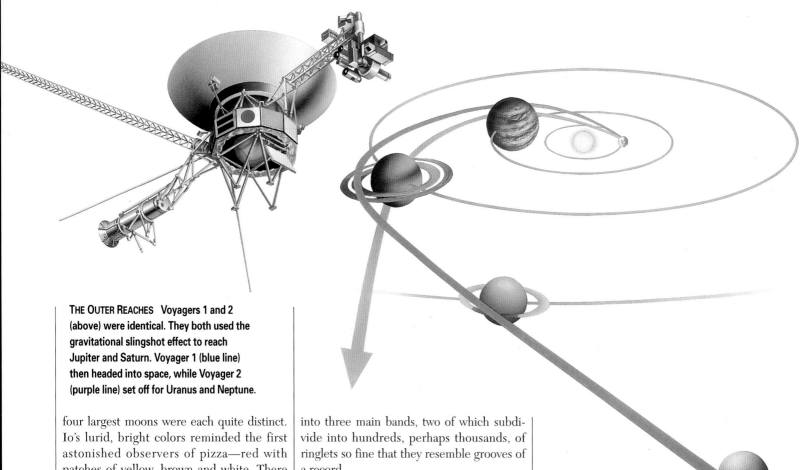

THE OUTER REACHES Voyagers 1 and 2 (above) were identical. They both used the gravitational slingshot effect to reach Jupiter and Saturn. Voyager 1 (blue line) then headed into space, while Voyager 2 (purple line) set off for Uranus and Neptune.

four largest moons were each quite distinct. Io's lurid, bright colors reminded the first astonished observers of pizza—red with patches of yellow, brown and white. There were also volcanoes, eight of them actually erupting—the first examples of active extraterrestrial volcanism.

The two Voyagers revealed the worlds beyond Jupiter. Saturn's giant moon, Titan, was found to have an atmosphere 50 percent denser than the Earth's. Though it turned out that all the gaseous giants have rings, Saturn's rings emerged as unique in their complexity and beauty. As the two Voyagers passed close by, transmitting data 100 times as fast as the Pioneers, the rings were seen to consist of millions of icy fragments, ranging from particles less than one ten-thousandth of an inch in size to boulders as big as a house, all in regular orbit around Saturn's equator.

Though each is only a few miles thick at most, the main rings stretch out over 44,000 miles. And since their orbital speeds vary, they divide

into three main bands, two of which subdivide into hundreds, perhaps thousands, of ringlets so fine that they resemble grooves of a record.

No one is yet sure how the rings were formed. For the most part, they are so close

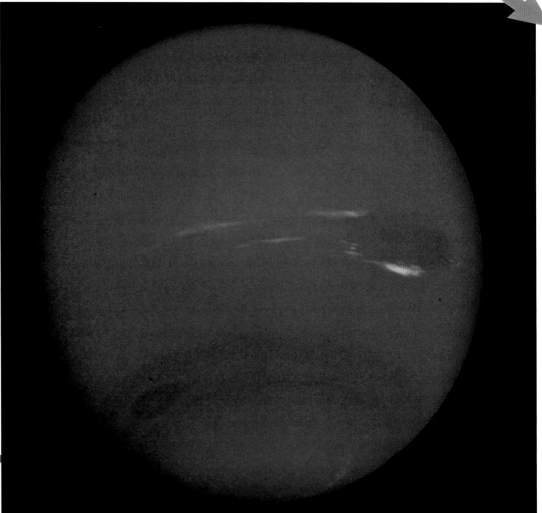

BLUE PLANET Neptune's methane-rich atmosphere is marked by high clouds that are prey to violent winds. A semipermanent storm forms a vortex known as the Great Dark Spot.

to Saturn that anything like a moon would be torn apart by the planet's intense gravitational field. So perhaps they are the remains of one or more captured moons that have been destroyed. On the other hand, perhaps they are made of primordial particles that could never have formed moons. Astro-

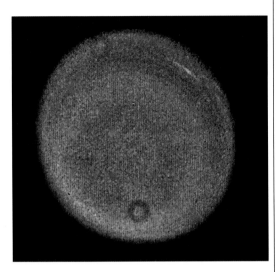

ATMOSPHERIC GASES   As Voyager 2 revealed in 1986, Uranus has an atmosphere rich in ammonia and methane. Like Saturn, it has rings, though they are much harder to see.

nomers hope that the solution to the mystery will cast further light on the way the planets themselves were formed.

The outer limits

Voyager 2 approached Uranus in 1986, exactly on schedule. Uranus turned out to be almost as fascinating a planet as Jupiter. While still a month away, Voyager spotted a tiny new satellite, named Puck; then two others, Cordelia and Ophelia. In all, Voyager discovered ten previously unknown little moons, which were all named after Shakespearean characters. Closer in, Voyager recorded clouds of ammonia with a delicately banded structure, moving with winds that blow in the same direction as Uranus's spin. The planet's tenuous rings are dark. Its magnetic field is large, and oddly positioned—offset by 60 degrees from its geographical poles.

Neptune, too far away to be seen by the naked eye, was known to have two moons, Triton and Nereid. In 1989, Voyager spotted six more, all small and rugged. It also sent back pictures of the planet's three rings and

A GALLERY OF MOONS

One of the most startling discoveries made by the four probes to the outer Solar System in the 1970s was that so many of the moons—58 are now known—are as individual as the planets they orbit. This is a portfolio of the most intriguing moons photographed by Voyager 2, with the Earth's Moon for comparison.

EARTH'S MOON

Our own Moon rotates at the same rate as it orbits the Earth, so it always presents the same face toward the Earth. It has no atmosphere and no active volcanoes, and is some 2,100 miles in diameter.

JUPITER'S MOONS

Io
Positioned close to its giant parent, Io's crust is continually being flexed and buckled. This creates internal heat that escapes through volcanoes, producing lava lakes and a tenuous atmosphere of sodium.

Europa
Europa consists of a rocky core covered with ice. Its craterless veneer is as smooth as glass, but cracked and ridged by shrinkage. Some astronomers have suggested that beneath the ice may lie an ocean of ordinary water.

Ganymede
Larger than Mercury, Ganymede has been icy for eons, but is pockmarked by craters and patterned by primordial cracks.

Callisto
Callisto looks something like the Earth's Moon, and is just as inert, but its impact craters are shallow because it, too, is covered in ice and the surface does not fracture on impact. One massive collision blasted a shallow basin 375 miles across, throwing up waves of molten rock, which eventually solidified to form a series of concentric rings, like the circles on a target around the bull's-eye.

SATURN'S MOONS

Titan
The largest of Saturn's moons, Titan has a thick atmosphere consisting mainly of nitrogen, which conceals its surface.

Iapetus
Like our own Moon, Iapetus's rotation matches its orbit, so the same side is always facing the planet.

Enceladus
Enceladus has a surface of grooves and fractures, a sign of recent stress.

URANUS'S MOONS

Miranda
The smallest of Uranus's moons, and the nearest to its mother planet, Miranda has a variety of different terrains, including grooved regions and a 3-mile high cliff.

Oberon
Oberon has a dense cover of impact craters and at least one high mountain, possibly a volcano.

Titania
Titania has an ancient surface scarred by craters and massive rift valleys up to 1,000 miles long.

Ariel
Ariel has valleys and craters, but they are lighter and probably more recent than those of Titania.

Umbriel
Umbriel is very dark, perhaps coated with material from an impact, but the surface has one light-colored impact crater.

NEPTUNE'S MOONS

Triton
Neptune's largest moon has a tenuous atmosphere, few craters, and no mountains or valleys. It has geysers of liquid nitrogen and lakes of solidly frozen nitrogen.

ENCOUNTER WITH A COMET

In the 1970s, scientists realized they were about to be presented with a unique opportunity. The best-known comet of all, Halley's comet, was on its way back after a 76-year absence. It was spotted in 1984, and would swing around the Sun in 1986. Several countries planned missions to investigate the comet, including the United States, Japan and Russia. The most ambitious would be sent by the newly formed European Space Agency. Its probe, named Giotto after the Italian artist who had painted Halley's comet into a picture in 1304, was launched by France's Ariane rocket.

The probes would examine an object whose origin was mysterious. When the Solar System formed, there remained a vast mass of unused material—dust particles, rocks, drifting gas. The heavier bodies became asteroids, while lightweight material was blasted into the outer reaches of the Solar System. There, according to theory, the gas and dust evolved into fluffy objects that every now and then were dragged toward the Sun, acquiring the typical cometary "tails" of dust as the Sun's rays batter them.

Astronomers and engineers were eager to see a comet close up, to investigate a number of mysteries about its structure. Giotto did not disappoint them. The 2,112-pound craft carried many experiments and a color camera, all shielded against dust and ice particles. This protection was vital as Giotto would be meeting Halley head on, with a combined speed of 150,000 mph, enough to slam a particle of dust through 3 inches of aluminum. At its closest, Giotto whipped through Halley's tail only 370 miles from the nucleus. Unfortunately, a few seconds previously Giotto had taken a hit from a dust particle, which knocked out the camera. But before this had happened, the sight was astonishing. The comet's core was shaped like a giant peanut, and measured about 9 miles long and 5 miles across. Its exterior was carbon-rich, "dark as velvet" in the words of one astronomer, with jets bursting through its dusty, cratered surface on the Sunward side.

Giotto also discovered that Halley's tail included some surprisingly complex chemicals, such as sulphur and carbon compounds, which are basic constituents of life—proof that Earth shares a common origin with comets.

BRIGHT TAIL Sweeping in from beyond Neptune, Halley's comet swings closer to the Sun than Venus, streaming its tail of ancient stellar matter.

FROZEN PEANUT The Giotto probe revealed that the core of Halley's comet is a dark, frozen mass shaped like a giant peanut (below). Giotto was launched to intercept the comet as it traveled away from the Sun (below right).

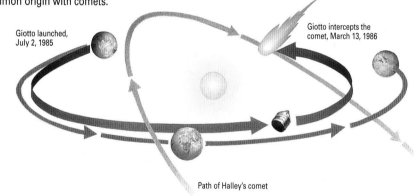

Giotto launched, July 2, 1985

Giotto intercepts the comet, March 13, 1986

Path of Halley's comet

beautiful images of Neptune's blue face, marked by a shadow named the Great Dark Spot, which was capped by white clouds. Winds blew over the planet at 700 mph, but the clouds above the Dark Spot did not move. The upper atmosphere consists of hydrogen and methane at a frigid temperature of –351° Fahrenheit, yet at its heart Neptune is warm.

Perhaps the most startling revelations were those to do with Neptune's biggest moon, Triton. Though smaller than our Moon, Triton has a slight atmosphere, but is so cold (–400° Fahrenheit) that it is partly covered by frozen nitrogen. This bitter world is volcanic, but its volcanism is unique in the Solar System: subsurface pressure forces up geysers of liquid nitrogen that explode out and then freeze into icy lakes.

As it passed Neptune at 60,000 mph, Voyager 2 had already traveled 4.5 billion miles since launch. Its signals were taking 4 hours to reach Earth, and were 36 times weaker than they had been from Jupiter. As it headed out of the Solar System in the early 1990s, the probe continued to broadcast, though it was then so far away that the Sun was no more than one star among many.

Nothing could have better justified NASA's faith in the scientific benefits of deep-space exploration. And nothing but Voyager 2's extraordinary pictures could have provided such a vital injection of publicity.

LIVING IN SPACE

EVER SINCE THE FIRST LONE ASTRONAUT ORBITED THE EARTH, THE ENDEAVOR TO EXTEND OUR POTENTIAL AS AN EXTRATERRESTRIAL SPECIES HAS CONTINUED. THE MOON LANDINGS SAW THE FIRST HUMAN STEPS ON A BODY BEYOND OUR PLANET, AND AS CONDITIONS IMPROVED, ASTRONAUTS SPENT LONGER AND LONGER IN SPACE. TODAY, THE SHUTTLE AND THE INTERNATIONAL SPACE STATION PROGRAM ARE WORKING TOWARD A PERMANENT FOOTHOLD IN SPACE.

COLONIZING THE NEW FRONTIER

WITH THE RACE FOR THE MOON OVER, THE SUPERPOWERS JOCKEYED FOR PRIDE OF PLACE IN EARTH ORBIT

As the Apollo program built up in the 1960s, von Braun's long-standing dream of a space station seemed about to come true. Eager to find further applications for the Saturn rocket, NASA began planning for a manned station. As soon as Apollo 11 was on its way, the decision became final. The station would be called Skylab, and it would be launched by a Saturn V, with three of the smaller Saturn I-Bs ferrying nine men into orbit to man it. Beyond this, NASA also wanted a larger, 12-man station—and more.

Soon after the launch of Apollo 11, the Space Task Group issued a wish list that would build on Kennedy's vision. Its cornerstone was a manned flight to Mars powered by a nuclear rocket, at a cost of $100 billion—five times the cost of landing a man on the Moon. In addition, NASA wanted a space shuttle to fly men back and forth into orbit; a space station in lunar orbit; more flights to the Moon; and a lunar base. America could have a man on Mars by 1983, said the report. Of course, it would take money; NASA's budget would have to rise to $8 billion a year. With high hopes, NASA went to the Office of Management and Budget, and asked for an immediate payment of $4.5 billion for 1971.

In response, NASA got a dose of reality. The budget was cut back to $3.4 billion. Taking account of inflation, this meant that NASA was receiving less than 50 percent of what it had in the mid-1960s. NASA administrators were forced to rethink their priorities for future manned flights.

The cost of sending a Saturn V into space was $650 per pound. NASA needed to lower the cost-per-launch. One way was to make bigger engines, which cost more to start with, but promised greater efficiency. Another possibility was to use strap-on boosters, like those attached to the Titan III, which would be used for the Viking and Voyager planetary missions. Such boosters could cut costs by a third. It would take time to evolve the right strategy. Meanwhile, a role still had to be found for the surviving Saturn rockets and Apollo spacecraft.

After Apollo, there were two possible directions for manned flights: a space station and the Shuttle. The Shuttle would take years to develop. Only the space station idea promised any hope of achievement in the short term. Besides, it also offered a final way of establishing American supremacy over Soviet ambitions in space.

A new goal for the Soviet Union

The Soviets had taken their defeat in the race to the Moon hard; almost until the end they thought they might win. The unmanned Zond lunar probes were supposed to have led to manned lunar orbital missions. But when Apollo 8 circled the Moon in December 1968, there was no point in trailing in the Americans' wake. One possibility remained: a flight direct to the Moon using a new "superbooster," of which the United States had only the vaguest inkling. This monster was rumored to be over 400 feet tall, with twice the mass of the Saturn V; in fact, it stood 312 feet high, and was about 10 percent more massive than Saturn.

This rocket could perhaps have come into its own if, as the Soviets hoped and believed, the American lunar landing was delayed until

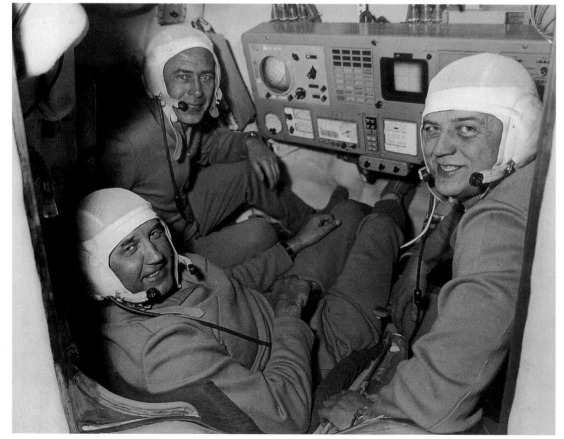

SPACE SUPERSTARS Dobrovolsky (lying), Patsayev (rear) and Volkov (right) pose for a cheery publicity shot while training for their doomed Soyuz 11 flight in June 1971.

1971 Salyut space station (U.S.S.R.) goes into orbit

1972 First Soviet-American arms limitation treaty

1973 Salyut 2 (U.S.S.R.) and Skylab (U.S.) both placed in orbit

1974 End of Skylab project; Salyut 4 placed in orbit

HONORED HEROES Soviet leader Leonid
Brezhnev heads the pallbearers carrying the
ashes of the three Soyuz 11 cosmonauts to
their final resting place in the Kremlin wall.

1972. As things turned out, they were over-taken by Apollo 11's success. At that stage, the Soviet obsession with secrecy paid off, and they simply announced that they had never intended to be in the race for the Moon. Official statements declared that the real future in space lay in space stations, not with wasteful manned missions to the Moon.

The plan was to establish a permanent space station. But an 18-day mission in 1970 revealed problems. The craft was made to spin, to create artificial gravity, and this had unexpected results. When they landed back on Earth the cosmonauts had to be carried away on stretchers because they were too weak to stand. It seemed that zero gravity was better than the artificial sort.

In April 1971, the Soviets put a 20-ton space station into orbit. It was called Salyut ("Salute") as a tribute to Yuri Gagarin, who had become an official Soviet hero, a virtual patron saint of the space venture. In June, three men were ferried up to the station in a Soyuz spacecraft, Soyuz 11. They were Georgi Dobrovolsky, the commander, and

engineers Vadim Volkov and Viktor Patsayev. The three men settled into their new home to terrific publicity. They performed nightly telecasts and quickly became stars, matching Gagarin and Tereshkova in popularity. This was a genuine first, and—according to Soviet propagandists—a truer sign of things to come than the journey to the Moon.

Then, on June 30, the twenty-fourth day of the mission, came a chilling official announcement. The three had been ordered home, and their capsule made a beautiful soft landing. But "the men were found in their seats without signs of life." Air had leaked out through a faulty valve while they were de-orbiting. The men, who were not wearing space suits, had suffocated in the few minutes before their capsule descended low enough for air to leak back in again.

There had been no warning: not for the three cheery and expert cosmonauts; nor for the officials who opened the capsule; nor for the public. The shock within the Soviet Union caused by the brief, grim announce-ment was immense, a trauma equivalent to

that in the United States caused by the assas-sination of President Kennedy. In addition to the tragedy of the death of three heroes, it was the end of a dream of restored national pride for the Soviet Union. The dead cosmo-nauts became the focus of nationwide grief. Hundreds of thousands filed past the open coffins as they lay in state in Moscow. In a moving tribute, an American astronaut, Tom Stafford, was one of the pallbearers.

Later, Soviet space experts provided an explanation of what had happened. The valve had been jolted open when the descent module separated from the rest of Soyuz in preparation for re-entry. The outrush of air had spun their capsule, confusing the men. When they realized what was happening, they struggled to close the valve manually. But in a fatal design error, equivalent to the flaws that had doomed the three Apollo 1 astronauts, it would have taken less than a minute for the air to escape—and twice as long to shut the valve manually.

Clearly, future cosmonauts would have to be protected by space suits. But that meant more equipment, if only to supply the suits with air. That in turn meant less room inside the capsule. In response, Soviet designers removed one of the bunks and turned Soyuz into a two-man craft. They also abandoned Salyut and began planning a new space station to be called Salyut 2.

Soviet policymakers found themselves backed into a corner. They had announced that they would win the race to establish a permanent space station, but the project remained dogged by problems. A Proton booster carrying a second station failed. Then, in 1973, six weeks before the United States was preparing to launch its Skylab space station, Salyut 2 reached orbit. This was a very different craft from Salyut 1, being a military station with a telescope. Its crew was supposed to take pictures, then send back the film using capsules equipped with their own retrorockets. However, no crew followed it up. Another spacecraft, code-named Kosmos-

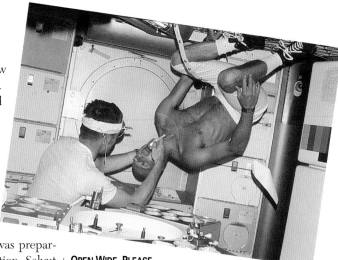

OPEN WIDE, PLEASE
Displaying a novel benefit of zero gravity, Skylab's commander, Charles Conrad, receives a dental examination without the need for a chair.

557, went up just four days prior to Skylab. Officially, it was a small scientific satellite. In fact, it later emerged that Kosmos-557 was another type of Salyut, and that the Soviet plan had been to pre-empt Skylab by establishing two space stations simultaneously. Once again, no crew followed because the two stations had unstable orbits. After a couple of weeks, they both fell back into the atmosphere, scattering flaming debris into the Pacific.

Success with Skylab

Meanwhile, American plans for Skylab had gone forward without major hitches. "Apollo gave us our Columbuses," one official said. "Now we need our Pilgrims. Skylab is the precursor of these." By coincidence, Skylab, at 90 tons, was about the same weight as the *Mayflower*, which had carried the Pilgrims to New England in 1620. Skylab was blasted off atop a Saturn V on May 14, 1973, heading for orbit, where it would receive its crew, who would follow in an Apollo craft.

A minute into the flight, disaster struck. As the rocket accelerated through the air, a cylindrical metal shield, designed to protect Skylab from dust and solar heat, tore away, taking one of the solar panels with it. In orbit, the second solar panel jammed shut. Though the telescope worked perfectly, deploying its own solar panels in a windmill configuration, those panels alone could not

BLOODIED BUT UNBOWED Skylab's mission badge (inset) shows the complete craft. But a near-disaster soon after takeoff left it orbiting with one solar panel missing, and a parasol in place of its lost solar shield.

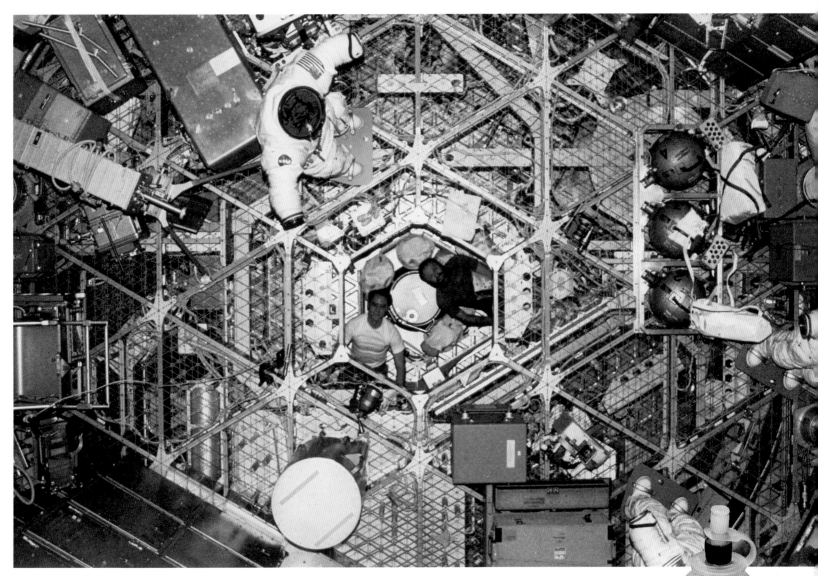

generate enough electricity to sustain Skylab and power the planned scientific experiments. Unshielded, and with virtually no power, Skylab's temperature soared to 125° Fahrenheit, threatening food, medicine, film and scientific experiments. The whole project seemed doomed.

On the ground, the crew's launch was put on hold. Again, as when Apollo 13 faced catastrophe, NASA fell back on a few dollars' worth of improvisation to save a project on which they had spent $2.5 billion. In the hopes that the Apollo crew could act as repairmen, they rigged up a parasol using fishing rods and metallic fabric. This would replace the missing protective panel.

In orbit 11 days later, crew members Charles Conrad, Paul Weitz and Joseph Kerwin maneuvered up to the stricken Skylab. Weitz leaned out of the hatch to try to lever the stuck solar panel open, without success. Then Conrad, veteran of Apollo 12, tried docking, and found the mechanism

wouldn't work. It took a 4-hour space walk—the longest to date—and yet more improvised repairs to correct things. Once inside Skylab, the crew pushed their parasol through the telescope's airlock, and flicked it open. Its shadow fell across the roasting metal, and at once the temperature began to fall. Soon, the three men could take off their cumbersome space suits.

Finally, after three days of rehearsal, Conrad and Kerwin donned their space suits again, ventured out, tied a rope to the tip of the stuck solar panel, and heaved. It swung free. "We see amps!" yelled an engineer in Houston, as Skylab began to regain power. The mission was saved. Over the next three weeks, the three men performed most of the planned science work.

Skylab was the first true home in orbit. With a length of 118 feet, and consisting of four sections, it was the size of a three-bedroom house. Its crews could sleep in their own com-

SPACE IN SPACE Until the launch of Skylab, spacecraft were either cramped or very cramped. Now, for the first time, astronauts had room to move with comfort. New equipment, such as this collapsible drink bottle (below and right) made long space trips more tolerable.

partments in anchored sleeping bags. There was a shower, and a toilet that overcame the problems of zero gravity with a suction pump. In a dining room, at a dining table, the crew could enjoy a range of food from the freezer, including lobster, steak and butterscotch pudding. Yet this first attempt to bring normality to space had its problems. Air bubbles built up in the food, with embarrassing consequences. As one later astronaut commented: "Passing gas 500 times a day is not a good way to go. The only redeeming feature is that everyone is passing the same amount." Vigorous daily exercise prevented the loss of muscle tone that would otherwise affect the astronauts in extended orbit.

Two-and-a-half months later, Conrad's Apollo 12 companion on the Moon, Alan Bean, led the second Skylab team aloft. A third team followed in November. These missions steadily extended the time spent in zero gravity from a month to 84 days. It seemed there was no limit in sight to the time that could be spent in orbit, or the range of work that could be done. The crew made detailed observations of both the Sun and Earth. In particular, their infrared images revealed geothermal energy sources in the western United States, water in drought-stricken regions of Africa, and possible sources of oil and copper.

By the time the Skylab project ended in February 1974, NASA had established that humans could survive and work in space for many weeks. The space station remained in orbit for another five years, before finally succumbing to the friction of gas molecules

and cosmic dust particles. These slowed it down, dropping it into ever lower orbits. Finally, in July 1979, Skylab fell into the Earth's atmosphere, and burned up off the coast of southeast Australia. It had been aloft for over six years, and had made a total of 34,981 orbits of the Earth.

Moving toward cooperation

Meanwhile, the Soviets were recovering from the Soyuz 11 tragedy. They were helped by being given access to the results of Skylab, as a prelude to a joint mission with NASA. In 1974-5, cosmonauts in Salyut 4 experimented with a small garden of peas to see how they fared in zero gravity. The plants lasted a few weeks, which suggested it might be possible to do better in the future.

On April 5, 1975, Salyut 4's second crew, Vasiliy Lazarev and Oleg Makarov, made a narrow escape when their third stage failed to separate, forcing an emergency return even as they approached orbit. One emergency led to another. For a few minutes, they feared they would land in China. "We have a treaty with China, don't we?" came a voice from the capsule before the radio blackout enforced by the heat of re-entry.

China, however, turned out to be the least of their problems. The capsule plunged far more steeply than any others had, subjecting the occupants to an appalling deceleration. Normally, cosmonauts suffered 4Gs—four times the force of gravity; 8 was considered

an emergency. This 15-minute descent increased their weight eighteenfold, to more than $1/2$ ton per man, sending the meter off the end of its scale. The parachutes eventually opened, and the capsule landed on a snow-covered mountainside. It rolled downhill, and was about to go over a precipice when the trailing parachute lines snagged on some stunted trees. Only when the rescuers arrived did the cosmonauts find out where they were—in the Altai mountains, near the Mongolian and Chinese borders, but still just inside the Soviet Union.

Salyut 4 had to wait another six weeks for a replacement crew to be launched. By the time they arrived, Salyut was beginning to deteriorate from the build-up of humidity. The windows were fogged over and the walls were green with mold. Still, the two men endured for 63 days, until they were allowed to come home. Salyut itself remained up for

MONSTER MOTOR

Among the possible successors to Saturn was a rocket with a giant engine being developed by Aerojet. This was a formidable beast. To ignite it required a second rocket, which itself produced 110 tons of thrust, and to ignite *that* demanded a third rocket generating 2 tons of thrust. In a test firing, this behemoth, which was in a pit with its engines pointing skyward, shot out flames that reached $1^1/2$ miles high.

another year-and-a-half before re-entering the atmosphere and burning up.

By the time Salyut 4's second crew returned to Earth, a turning point had been reached in terrestrial politics. The international mood that had inspired the space race had changed. The Cold War was almost dead, and *détente* was the new buzzword. Far from being rivals, the superpowers were now said to be "converging": two immense

economic powerhouses advancing toward wealth and happiness along different routes. In 1972, President Nixon had gone to Moscow, where he had signed the first nuclear nonproliferation treaty with the Soviet Union. In 1973, Soviet premier Alexei Kosygin visited the United States.

The superpowers would continue to maneuver for political advantage, but their rivalry was now limited to a few hot issues: cruise missiles and SS-20s in Europe, Israel's occupation of part of Egypt. Even when old Soviet-American rivalries soured relations again, few now believed that nuclear war would break out between the superpowers.

Planning a joint space venture

Nevertheless, both nations were left with the space programs that their rivalry had once produced. Both needed to find a good use for their hardware. They did so with an agreement for a joint Soviet-American space mission. The venture inspired euphoria and skepticism in equal measure. It was good

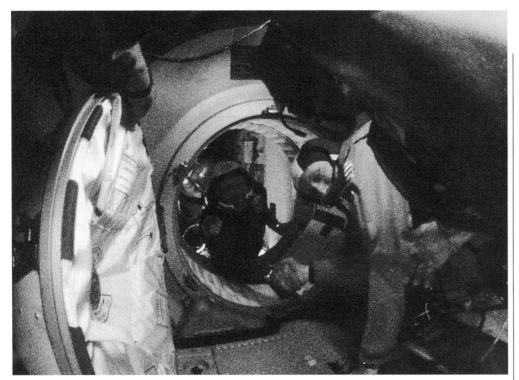

FRIENDS IN SPACE In a deliberate gesture of rivals' peace and cooperation, Tom Stafford and Alexei Leonov shake hands for the camera after the docking of the two craft that formed the Apollo-Soyuz Test Project (ASTP).

publicity for détente that Soviet and American astronauts were seen training in each other's countries, or at ceremonial functions, or hunting or skiing together. But part of the deal was an exchange of information. Since the United States was ahead in both rocketry and orbital space flights at the time, American analysts feared that the Russians would be the main beneficiaries. "Our arms around their shoulders and their hands in our pockets," was one official's comment.

The original idea for the joint venture was for an Apollo spacecraft to dock with a Salyut space station. As it happened, existing Salyuts could not take a second link-up in addition to their own Soyuz ferry craft, and more advanced versions would not be ready in time. The link-up would have to be between two spacecraft—an Apollo and a Soyuz, giving the mission the name Apollo-Soyuz Test Project, or ASTP.

On July 15, 1975, the Soviet Soyuz rose 138 miles into orbit. A few hours later, the Apollo went up. It maneuvered, approached and, with infinite care, docked, almost 52 hours after the Soyuz was launched. After a few more hours, the air pressure was equalized, the hatches opened, and TV viewers were treated to handshakes and hearty back-

slaps between men who had spent months training together and now knew each other as friends.

The celebrations were milked by both nations for every ounce of symbolic significance. The United States was eager to use publicity to confirm Moscow's commitment to détente. The Soviets wanted to set past failures behind them and be accepted as an equal in space exploration.

There were also real immediate benefits. The Americans used a satellite for the first time to communicate with Mission Control, and experienced a threefold improvement on their previous system. The Soviets used the mission to open a new control center in Moscow. Finally, the Apollo flight itself did useful work. The Soviets returned home after six days, leaving the three Americans in orbit for another four days, during which time they conducted experiments on processing materials in a special furnace.

In the end, though, the project was more show than substance, and within a few years the intentions of both sides proved illusory. Soviet leaders would not allow themselves to be dictated to. In 1979, they would invade Afghanistan, virtually redeclaring the Cold War. And Soviet space expertise would never match that of the United States. In practical

GLAD TO BE HOME Valentinovich Volyanov and Vitaly Zholobov look relieved to be back on Earth after being suddenly recalled from the faulty Salyut 5 on August 24, 1976.

terms Apollo-Soyuz led nowhere. Even the agreement to exchange information had no great significance, for the Apollo technology was already ten years old, and most of what could be revealed was already public knowledge. If the Soviets dreamed of a flood of top-security material, they were disappointed, for this was the last of the Saturn rockets, the last of the Apollo spacecraft, and no more Skylabs were planned.

Even so, for the Americans, Skylab and the Apollo-Soyuz mission provided a bridge to a new era. The days of massive launch vehicles—Saturn V and Korolev's SL-15— were now over. The immediate future of manned flight would be in Earth orbit. Skylab and Apollo-Soyuz created a role for all the remaining Apollos and the thousands of people who worked on them, preserving expertise, hardware and manufacturing capacity until it was time for the next great venture—the Shuttle.

In space to stay

Soviet achievements built a bridge to the future of a different kind, the significance of which would not emerge for many years. Though plagued by technical problems and

some outright failures, the Soviet Union's Salyut space stations continued to provide scientific research and surveillance for a decade after the 1975 Apollo-Soyuz mission. The Salyut missions also produced an ever-growing body of information on the ability of humans to adapt to long periods in space. Given their lower levels of safety and comfort, the Salyuts also honed a particularly Russian talent for enduring tough conditions and coping with crises.

Salyut 5, which went into orbit in June 1976, was another military station, probably designed for successive crews to keep an eye on military installations and maneuvers in the United States and China. In addition to several cameras, it included a 1,500-pound infrared telescope that straddled the center of the Salyut.

Staffing the new station led to problems. The first two crew members were recalled suddenly after seven weeks when the air-regeneration system developed a fault. In October 1976, a second two-man crew suffered a guidance-system failure, and headed for home at once. Below, it was night, and the Kazakh grassland, their landing zone, was blanketed by the autumn's first big snowstorm. As bad luck would have it, icy winds dumped them in one of the region's few large lakes, Lake Tengiz, which was not yet frozen. They were left bobbing on the waves, in the dark, with the temperature dropping fast. It took 6 hours to retrieve them. Rescuers were unable to attach lines, dinghies dropped by helicopter were blown off course and a helicopter failed to lift the 3-ton capsule. In the end, the helicopter dragged it to shore and across a mile of swampland to firm ground, where other helicopters could land. This epic rescue was not reported at the time. Officially, the crew was declared safe after "minor difficulties."

A third crew finally got into Salyut 5 in early 1977. They managed to repair it and to carry out some of the planned experiments. But they could not stay long. The station had been up for eight months and was nearing the end of its useful life. The crew left after two weeks. The station itself burned up six months later.

By then, Salyut 6 was in orbit. In early October 1977, its crew rose to meet it in Soyuz 25. Supposedly, they were to dock with Salyut as a "salute" to two historic events—Gagarin's orbital flight 20 years ear-

GUESTS ALOFT

In 1976, the United States signed an agreement with the European Space Agency, by which Europeans would build a laboratory for the planned Space Shuttle, and also contribute astronauts. This step into the international arena focused the minds of Soviet space administrators. Soon afterward, they announced their own "guest-cosmonaut" program.

It so happened there was room in the second seat of the Soyuz spacecraft being prepared to resupply the Salyut space station. It would make good political sense to offer this to a non-Soviet cosmonaut. Soviet propagandists could use the flights to assert the equality of all members of the Soviet world.

Two pilots were invited from each of several socialist states—Czechoslovakia, Poland, East Germany, Bulgaria, Hungary, Mongolia, Cuba, Romania and Vietnam. Their role would be kept simple, although they required a year's rigorous training. The prime requirement was that they were jet pilots; the second was that they had to be good Communist Party men. The first guest cosmonaut, Vladmir Remek, was the son of Czechoslovakia's deputy defense minister. One of the Poles, Miroslaw Hermaszewski, was the brother of an air force general. Their credentials were clearly impeccable.

As propaganda, the program worked well. Though none of the guest cosmonauts contributed anything much to practical astronautics, they became heroes in their own countries. And the Soviets were able to claim yet more space firsts: the first Asian in space (the Vietnamese Pham Tuan) and the first black cosmonaut (Arnaldo Tamayo-Mendez of Cuba). In the mid-1980s, America's Shuttle made commuting to orbit routine, but for almost a decade the Soviet Union justified its claim to be the pioneer of internationalism.

SOVIET "FIRST" Arnaldo Tamayo-Mendez (left) became the first black man in space when he flew in Soyuz 38 with the U.S.S.R.'s Yuri Romanenko.

lier and the October Revolution of 1917. But docking failed, presumably because of a fault in the Salyut. Low on fuel, and disappointed, the two crewmen returned to Earth.

This presented an embarrassing problem. The newest, most sophisticated Salyut, with its fancy stellar navigation system, would be virtually useless if the docking mechanism of the main port had failed. A veteran of Salyut 4, Georgi Grechko, who had also helped to design the docking mechanism, was hastily called in as a repairman. In orbit, in December 1977, Grechko became the first

Russian in nine years to walk in space as he attempted to find out what was wrong with Salyut's docking mechanism.

As it turned out, there was nothing wrong with the dock, but the cosmonauts' relief almost led to a tragedy. Grechko turned to admire the view as the Pacific wheeled below him. His colleague, Yuri Romanenko, inside the airlock, stuck his head out too, and then with a little push allowed himself to drift outside, his safety line trailing behind him. Suddenly, both men saw a sight to stop the heart—Romanenko had forgotten to attach the end of his line. Within seconds, he would have been out of reach, drifting away to suffocate hours later as he drained his air supply. As Grechko revealed later, he seized the line and hauled his partner to safety.

A Soviet endurance record

After that dramatic start, the two men settled down to set a new space endurance record, while working at their science projects. Unexpected problems constantly arose. They experimented with a little garden, but the seedlings failed to thrive. An experiment with fruit flies, which were to be watched for possible genetic changes, went wrong when the container cracked, allowing the flies free to roam for days until they were vacuumed up.

The men also reported experiencing heightened and distorted sensations. The canned ham tasted too salty. The station's noises—the teletype machine, ventilation

EXERCISE—THE KEY TO LONG-TERM SPACE SURVIVAL

Human health depends on working against the pull of gravity, which keeps muscles and bones strong, and the heart pounding. Without gravity to fight against, muscles, bones and hearts all weaken. Other changes follow. Rather than pooling in the legs, blood is distributed evenly throughout the body, with more reaching the heart. To rectify this apparent excess, the body cuts back both on vital chemicals and red blood cells, with the result that it cannot cope with normal amounts of stress. Headaches increase, as does fluid intake. The best way to counteract this in space is by exercise—up to 2 hours of it a day.

Since the start of the Salyut and Skylab missions, crews have worked out using three main devices: on treadmills; with bungee cords, which provide resistance; and with exercise cycles. But exercise produces its own problems. For instance, if sweat is absorbed in clothes, it produces rank smells in a day. If the astronaut exercises without clothing, he finds that sweat does not evaporate or run anywhere. It accumulates on the skin in slimy puddles, or drifts away in minute globules.

The crew of Salyut 6 found their workouts so unappealing that they skipped two-thirds of their exercise routines, and as a result were too weak to walk when they arrived back on Earth. Later space crews would not make the same mistake.

TROUBLESHOOTERS IN SPACE Yuri Romanenko (left) and Georgi Grechko overcame a string of technical and psychological problems during their record 96 days aloft in Salyut 6.

fans, creaks as the hull cooled and heated—got on their nerves. Facing these stresses over extended periods, the men themselves were the main experimental objects. They became the basis of a new science—space medicine—in which Soviet scientists acquired particular expertise. Each physical problem, once identified, could be countered; and the body itself learns to adapt to its environment after about a month.

A number of psychological problems also arose, especially those posed by isolation and enforced togetherness, and the two crew members became supremely alert to signs of trouble. As Grechko explained later, the key element was joint responsibility. "We decided that neither of us would press the other, never give orders." Both men learned to forestall trouble by checking on the other's mood. Both would start on unpleasant tasks whenever they saw the need. "If you see the toilet filter needs changing, you put down your interesting job and start the dirty one. And he does the same." In addition, Soviet Mission Control used a psychological support group, which made sure that family, friends, colleagues and specialists were on hand to provide support and feedback.

Every month, the crew had visits bringing letters, fresh food, fuel and specially requested items. Each was a first of its kind. In January 1978, two Russian cosmonauts became the first to dock with a manned space station. Nine days later came the first docking of a robot craft, Progress 1 (it was later packed with garbage and sent off to

burn up in the atmosphere). And in February, another two-man mission included the first non-American, non-Soviet astronaut—a Czech, Vladimr Remek.

The first visitors arrived with a serious emotional dilemma. One of the new arrivals murmured to Romanenko that Grechko's father had just died. Should he be told? If so, it might threaten his emotional stability and even undermine the mission. Romanenko decided against telling him. Too much depended on Grechko's peace of mind. But Romanenko insisted on taking responsibility; when they were safe on Earth, he would be the one to break the news. The secret was kept, and later Grechko was the first to acknowledge his colleague's wisdom.

Grechko and Romanenko returned to Earth on March 16, 1978, after 96 days aloft. When they landed they could not walk, having done only one-third of the prescribed exercises, and their return to gravity made them susceptible to dizziness and nausea. They, more than the Skylab crews, demonstrated what was possible in orbit. They had been resupplied by space "ferries." They had shown the difficulties of space living, and the solutions. Moreover, they had maintained an emotional stability and mutual support that was as crucial as it was extraordinary. Men were in space to stay.

THE SHUTTLE

AS THE POLITICAL MOTIVES FOR FURTHER SPACE RESEARCH FADED AWAY, A NEW ERA OF MANNED SPACE FLIGHT SUDDENLY DAWNED

It had always been acknowledged that rockets were ludicrously wasteful space vehicles. The first two stages of a Saturn V ended up in the ocean, while the third stage and service module were dumped in space—all those engines, computers, safety devices and life-support systems gone forever. Taking a three-stage rocket to the Moon was like crossing the Atlantic by ocean liner, ejecting bits along the way and then returning in the lifeboat. Such costs might be bearable temporarily when testing new technologies, or when balanced against profit from the scientific discoveries of a mission. But no nation could afford to throw away a $400 million Saturn with every launch indefinitely. Real savings could only come from re-usable craft in which the only element lost would be the fuel.

Hence the concept of the Space Shuttle—a re-usable rocket plane that would fly into and out of orbit. Such a vehicle had been in the dreams of space-flight engineers from the start. A space plane, a natural extension of the X-series of high-flying planes developed in the

1950s, was at the top of NASA's wish list as Apollo hit its stride, and was still alive after the cost cuts of 1970 and 1971.

At first, there were two possible solutions. One consisted of a two-stage rocket, both stages of which would have wings and pilots. The first stage would carry the smaller second stage to near-orbit, and both would return. The second solution involved massive fuel tanks that dropped away and burned up, leaving a modest, plane-sized winged rocket free to return from space to a runway. In both cases, the basic concept gradually grew in size.

NASA initially considered a 45-foot bay holding payloads up to about 11 tons. The Air Force, the prime end-user, wanted 60-foot holds taking up to 30 tons. And the Air Force's preferences mattered, because without its backing NASA might get nowhere.

On this basis, NASA started to refine designs. Both possible approaches foresaw a breathtaking new machine, in two versions. Both would have winged first stages some 250 feet long, a little larger than a Boeing jumbo jet. The Orbiter would be some 200 feet long. Both stages would have pilots to fly them back to Earth. The whole thing would weigh 2,500 tons at liftoff, comparable in size to a fully fueled Saturn-Apollo. This would, in effect, be a rocket plane picking up from where the X-20, the X-15's successor, left off—but 75 times as heavy.

Yet NASA suggested that each Shuttle would fly for $4.6 million, which, given the fact that all the hardware was re-usable, would be one-tenth the cost of an Apollo. This equated to a launch cost of $70 per pound—once the development costs of the

HIGH FLIER The X-15 rocket plane, which could penetrate inner space, provided inspiration for the idea of a re-usable space vehicle. The young pilot in this 1960 photograph is Neil Armstrong.

HEAT SHIELD Not the tiniest gap can be allowed between the protective tiles on the Shuttle's hull. Over 30,000 of them have to be fitted by hand, with painstaking accuracy.

project had been covered. And those were approaching $10 billion.

To skeptics in the Office of Management and Budget, whose job it was to check these figures, NASA's plans seemed over the top. "Unrealistic," commented one top official. On an annual budget of less than $3.5 billion, NASA couldn't afford it. They were told they would have to budget for development costs of only $1 billion a year.

James Fletcher, NASA's new boss, realized that the whole project was at risk if NASA did not adapt. As Senator Walter Mondale argued as part of a campaign to kill off the Shuttle project, this was a "classic case of a program and agency in search of a mission." The answer was to go for cheaper development costs and higher running costs. From this grew the design that would dictate the eventual look of the Shuttle as it is today.

The key element was to put most of the fuel on the outside of the Orbiter, thus reducing the Orbiter's size. There would be one big external tank and two re-usable strap-on boosters. The price tag began to look more acceptable: development costs of just over $5 billion, half the original estimate. Launch costs per pound would be relatively high at

$160 (at least so NASA claimed), but by then the Shuttle would have a virtual monopoly of space transportation and would therefore be able to pay its way. The arguments were good enough. Nixon liked the idea, and in 1972 Congress gave its approval as well.

Developing the Shuttle seemed a natural progression from Apollo. The main contractors were those that had worked on the lunar mission. Cape Canaveral, with its Saturn V facilities, could become the Shuttle's launch site. Rockwell International, with all its Apollo experience, signed the Shuttle contract, worth $2.6 billion. Rocketdyne would make the Shuttle's main engine. The only team member missing was the man who started it all, Wernher von Braun. He had set his sights on schemes that were too grand for these cost-conscious times, and resigned in 1972 to go into private industry.

New engines, new problems

The original plans called for the Shuttle to fly in 1978. But the Orbiter's new engine—the so-called SSME (Space Shuttle Main Engine)—proved problematic. There were to be three of them. Each had to produce more power in a more compact form than any previous engine. To deliver the required flow of liquid hydrogen, its fuel turbopump, the size of a large outboard motor, had to develop 76,000 horsepower, more than many early transatlantic liners. Problems multi-

plied, in part because of the commitment to an approach that had worked well for Apollo—"all-up testing," which meant testing only the complete engine, rather than its elements separately. But Apollo had been constructed on the basis of years of previous experience with other rockets, so this form of testing worked. The SSME started from scratch, and the testing didn't work. One fire followed another with depressing regularity.

Another problem was thermal protection of the Orbiter on re-entry. Apollo's shield had been designed to burn away. The Orbiter needed a long-lasting shield that would radiate the heat. The solution was provided by ceramic-coated, silica-fiber tiles, fitted to perfection. There would be some 31,000 tiles, 6 inches square, and it took one person three weeks to install four of them. When the first Orbiter traveled to Cape Canaveral in March 1979, 2,000 workers went with it to install the last 10,000 tiles—only to see many of them pulled off by their own colleagues, who had to prove they were fixed firmly enough to resist the fearsome heat and blast of re-entry.

Nevertheless, even as the launch date slipped, the Shuttle remained popular with legislators, for it had acquired a military role. President Carter was trying to agree on more disarmament with the Soviet Union, and the Shuttle offered a way to launch the satellites needed to verify the mothballing of Soviet bombers and missiles. The first flight was scheduled for March 1981. Further difficul-

DYNA-SOAR: THE SPACE SHUTTLE'S LITTLE DADDY

The first re-usable space-plane design emerged in the late 1950s as a follow-up to the X-15, an American rocket plane that could fly halfway to orbit and reach 2,300 mph. Designated the X-20, this 6-ton reconnaissance plane, for which Boeing won a contract in 1959, would ride a Titan III rocket into orbit over its target area. It would then be released to glide back to Earth. The maneuver, known as "dynamic ascent and soaring," gave the project its name: Dyna-Soar. It went far enough for six test pilots to start training, but in 1963 American Defense Secretary Robert McNamara decided its job could better be done by an orbiting laboratory—which was in turn killed off in 1969 as unmanned satellite reconnaissance improved.

ties delayed blastoff until April 10, when the countdown halted with 9 minutes to go after a computer malfunctioned. Two days later, a million onlookers, and the world, watched as the Shuttle rose on a pillar of flame from billowing clouds of steam and smoke, showing at last the means by which men and women could commute into space.

Each $250 million Orbiter is 122 feet long, about the size of a medium airplane. A two-story crew compartment has seating for seven people (commander, pilot and various mission specialists). The 60-foot cargo bay is the size of a small railroad car. On launch, the Shuttle's engines take their power from a disposable fuel tank the size of a grain silo—154 feet long and almost 28 feet across. Yet most of the launch thrust comes from two solid-fuel boosters, which are manufactured in sections and later assembled at the Cape. These two solid-fuel rocket

AFRO-AMERICAN ASTRONAUT When the first Shuttle was launched, Guion Bluford was already in training for a later flight. In 1983, Bluford became the first African-American in space aboard the Shuttle *Challenger*.

boosters fall away after 2 minutes. They parachute into the Atlantic for recovery and re-use. The Shuttle powers on upward, draining the external tank, which is then jettisoned to fall into the Indian Ocean, leaving the Shuttle on its own in space. At the end of its mission, the Shuttle lands in California, and is flown back to its launch site perched on top of a specially adapted Boeing 747.

That first flight was a bit of a gamble. The Shuttle's engines had never operated in flight, and no flight had actually tested whether the thermal tiles would stay in place. The pilot, John Young, veteran of two Geminis and two Apollos, was one of the few survivors from the early days. He had actually been on the Moon when the Shuttle was approved by Congress in 1972. Now, at the age of 50, he was

SPACE COMMUTERS Veteran astronaut John Young (left) and his youthful co-pilot, Robert Crippen, pose with a model of the Space Shuttle *Columbia* before its successful maiden flight in April 1981.

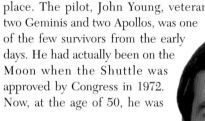

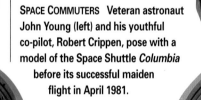

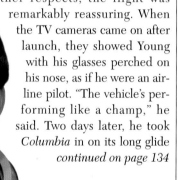

taking the Shuttle *Columbia* on its maiden flight, with Robert Crippen as his co-pilot.

On takeoff, 16 thermal tiles fell off, causing some anxiety at Mission Control. In other respects, the flight was remarkably reassuring. When the TV cameras came on after launch, they showed Young with his glasses perched on his nose, as if he were an airline pilot. "The vehicle's performing like a champ," he said. Two days later, he took *Columbia* in on its long glide *continued on page 134*

1981 Maiden flight of the first Shuttle, *Columbia*

1983 Maiden flight of Shuttle *Challenger*

1984 Maiden flight of Shuttle *Discovery*

1985 Maiden flight of Shuttle *Atlantis*

1986 Shuttle *Challenger* blows up, killing seven crew

THE SPACE SHUTTLE

EVERY SHUTTLE MISSION FOLLOWS A NOW FAMILIAR COURSE FROM LAUNCH TO EARTH ORBIT TO LANDING, THEN BACK TO BASE TO PREPARE FOR THE NEXT LAUNCH

The spacecraft that is now commonly called the Shuttle is known to NASA as the Orbiter, the term "Shuttle" being used for the whole assembly, including solid rocket boosters and external fuel tank. "Orbiter" is a descriptive name because it indicates precisely the limits of the craft's design and purpose: it is intended to orbit the Earth at an altitude of 115 to 690 miles. It cannot enter deep space and will never fly to the Moon or another planet, in its present form at least. Even so, the Orbiter is in many ways the most successful space vehicle yet conceived. By the beginning of 1999, NASA's five Orbiters had flown 93 missions, and with the notable exception of the *Challenger* disaster, their flights have become so routine that they merit only the briefest mention in the news, if they are noticed at all. Yet each Orbiter's space flight is only one stage in a complex, ongoing process that involves launching, flying, landing, recycling and then transporting back across America for its next launch.

In the early days of the Shuttle program, Orbiters were launched from Cape Canaveral, and landed there as well. In recent years, Shuttles have consistently taken off from Cape Canaveral, and landed at Edwards Air Force Base, in California.

A launch can take place during the day or night, depending on the Shuttle's mission and the best time for achieving the correct trajectory. Known as the "launch window," this can be a few minutes or a couple of hours in duration. If all proceeds according to plan, the Orbiter's three liquid-fuel engines power it on its way with 495 tons of thrust, assisted by two solid-fuel booster rockets, which provide 1,183 tons of thrust each. These fall away at an altitude of 28 miles, followed by the huge external fuel tank at about 68 miles. The Orbiter's two maneuvering engines then propel it upward to its desired orbital altitude, where work can begin.

The crew then busy themselves with the many different tasks that each flight might entail: launching a satellite; capturing a broken satellite; conducting scientific experiments; rendezvousing with a space station and transferring crew and supplies; testing new equipment. There is no "typical" mission; each one has its own agenda, lasting from a few days to over three weeks.

With the astronauts' space duties completed, the pilot brings the Orbiter in on its long glide through the atmosphere to land at Edwards Air Force Base, north of Los Angeles. Here the Orbiter is meticulously examined for any signs of wear and tear, before being loaded onto the back of a modified 747 jet, which transports it back to Cape Canaveral. It is then refurbished and loaded with its next cargo before being taken to the Vehicle Assembly Building, where it is attached to its fuel tank and booster rockets. The complete Shuttle is then ready for its next launch.

OFF AGAIN The Space Shuttle lifts off from Cape Canaveral, after being refurbished and fitted with a new fuel tank and recycled solid rocket boosters. The main fuel tank was originally painted white, but since the third Shuttle mission in 1982 it has been left unpainted, with a weight savings of 600 pounds.

CHERRY PICKER One of the Shuttle's most versatile pieces of equipment is its Remote Manipulator System (RMS), a mechanical arm. It has been used for lifting and manipulating the Shuttle's cargo, for retrieving satellites in orbit, and as a work platform for astronauts (as here). In order not to float off into space, the astronaut has his feet anchored to a small platform known as the RMS's mobile foot restraint.

PERFECT LANDING After entering the atmosphere at more than 17,500 mph, the Shuttle executes a series of maneuvers to slow it down, before its final landing approach at some 220 mph. Unable to fire its engines in reverse, like a jet aircraft, the Shuttle deploys a small drag chute that helps it roll to a stop without burning out its brakes.

HEADING FOR THE CAPE The 747-Shuttle duo takes off from California on its way back to the Kennedy Space Center at Cape Canaveral. There the Shuttle will be prepared for a new launch, and a whole new cycle will begin again.

RIDING PIGGYBACK The Shuttle is mounted on top of a specially modified 747 to deliver it back to Cape Canaveral. A special cone covers the Shuttle's exhaust nozzles to improve the aerodynamics of this very awkward combination.

FLYING FREE Robert Stewart, one of five crew members aboard the Shuttle *Challenger* in February 1984, tests the Manned Maneuvering Unit (MMU), which liberates astronauts from being tethered to the Shuttle. Nine months later, Dale Gardner used his MMU to secure and rescue the ailing Westar VI satellite (left).

from orbit to a perfect landing at Edwards Air Force Base. *Columbia* survived the loss of its 16 protective tiles.

It was the beginning of a few years in which NASA seemed to have rediscovered its way. Three other Orbiters—*Challenger*, *Discovery* and *Atlantis*—rolled off the production lines. Between them, these four amassed 24 missions between 1981 and early 1986. There were problems: for one thing, the Shuttle orbited at a maximum altitude of 690 miles, and most satellites needed to be higher, which meant they had to carry their own boosters. Overall, however, the first flights showed how effective the new vehicle could be. The range of crew members grew: Dr. Sally Ride became the first American woman in space, Guion Bluford the first African-American astronaut. Later missions carried a Canadian, a Frenchman, a Mexican, a Saudi prince and a few politicians. In all, during those first five years, more than 100 men and six women rode into orbit, performing experiments, monitoring their own bodies, testing equipment, setting new satellites in orbit, and repairing or recovering old and damaged ones.

One major development for space walkers was a move away from the systems of tethers and umbilical cords that supplied oxygen and radio links. In February 1984, astronauts were given freedom to roam with the Manned Maneuvering Unit (MMU). Using hand controls, the driver, or occupier, could twist, turn and fly his armchair-like space-car in any direction by controlling 24 tiny nitrogen thrusters.

With this device, crewmen could make repairs more effectively. The first to be tackled was a stricken NASA observatory, the Solar Maximum satellite, which had been launched in 1980 to study the Sun. Now it needed some replacement electronics. On the eleventh Shuttle flight in April 1984, Bob Crippen, now commander, stationed *Challenger* within 200 feet of the crippled satellite, but George Nelson in his MMU was unable to wrestle it aboard. The next day, *Challenger's* mechanical arm caught the satellite and brought it into the cargo bay, where crewmen made the repairs.

Toward the end of the year, crewmen in MMUs manhandled two communications satellites on-board, ready for refurbishment on Earth. Whatever the earlier problems and delays, it seemed the Shuttle would be able to justify itself.

Advancing the cause of science

Meanwhile, NASA was eager to pursue the other strand of its vision for a manned future in space: a space station, a concept that was now intimately bound up with the Shuttle. As the Shuttle was built, designers in Houston proposed a Space Operations Center, an array of modules brought up by

SPACELAB

In November 1983, the Space Shuttle went international, carrying the European research module, Spacelab, into orbit. The lab, designed by the European Space Research Organization (forerunner of the European Space Agency), carried 77 experiments, which would have been utterly beyond an unmanned mission. On board was the first non-American crew member, a German, Ulf Merbold, whose job was to supervise Germany's contributions to the mission. The six-man team stayed up for ten days, working around the clock in two shifts.

In 1985, the second Spacelab (actually designated Spacelab 3) carried a crew of seven, three of them over 50 years old. Later that year, a third Spacelab became the Shuttle's first chartered flight, with all 75 experiments provided by West Germany. These experiments focused on the behavior of fluids, crystals and metals in zero gravity.

Later Spacelab missions in the 1990s studied the effect of weightlessness on humans and other, smaller living organisms, such as rats and jellyfish. The success of Spacelab showed that scientists, as well as astronauts, could do useful work in space.

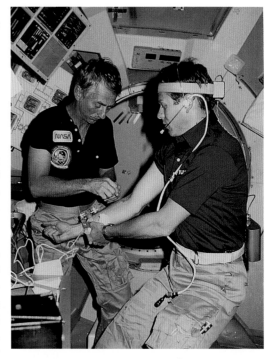

BLOOD SAMPLE On board the European Spacelab, a crew member draws blood from another for medical analysis. As the mission badge (inset) shows, Spacelab was designed to have ample working room for its crew. It also fitted the Shuttle's hold perfectly (below).

the Shuttle; the main purpose of this new space station would be to assemble and maintain satellites in orbit. With encouragement from NASA's new boss, James Beggs, studies continued, in the hopes of winning support from the president, Ronald Reagan, but he backed away from the idea. Instead, it became clear that satellites were becoming ever more reliable. If the space station had to rely on satellite repairs for its justification, it would not be in business for long.

NASA returned with another justification: zero-gravity research and development in two fields—medicine and electronics—both of which might offer significant returns. These proposals received a dramatic boost from the experiments conducted in several early Shuttle flights, but at the time there was still skepticism. Senator William Prox-

mire, one of NASA's most severe critics, told space-station backers they only wanted a station "to keep your centers open and your people employed." But the success of the Shuttle swung Reagan behind the station.

In January 1984, Reagan's State of the Union speech included a ringing endorsement in the Kennedy tradition: "We can reach for greatness again. We can follow our dreams to distant stars, living and working in space for peaceful economic and scientific gain. Tonight, I am directing NASA to develop a permanently manned space station, and to do it within a decade."

In August 1984, Shuttle crew members showed that it was possible to unfold a 102-foot solar array, which would provide power for a space station. A year later, using scaffolding tubes, other Shuttle astronauts prac-

ticed the techniques that would be used to construct a permanent station in space.

To the outside world, NASA was riding high; but there were problems. First, the Shuttle's commercial base was weak. France's Ariane came on line at just the time when NASA's ambitions had exposed it to competition. With the Shuttle supposedly approaching completion, NASA declared that the days of the expendable booster were over, as far as it was concerned. As of 1983, established workhorses such as Titan II, Atlas-Agena and Thor were to be transferred to the private sector. There, they faced a tough future competing with the Shuttle and Ariane, which were both heavily subsidized.

That should have made marketing the Shuttle easier for NASA. But NASA was committed to recovering costs, and had to seek payment of $71 million for the use of its cargo bay in full, in advance. Ariane, on the other hand, asked only for a deposit, with later payments spread over years following the flight. The result was that NASA, though increasingly monopolizing the American market, was still undercut by Ariane.

A further blow came when the giant Ortho Pharmaceuticals, which had been a major backer of the Shuttle as a zero-gravity laboratory for the production of an anti-anemia protein, pulled out. The company had adopted a new process that was not susceptible to gravity.

Worse, the Shuttle's technical foundations were proving shaky. The engines were supposed to make 55 flights before needing an overhaul, but this was not the case in practice. The main problem lay with the highly sophisticated, $36 million turbopumps, which needed new turbine blades every three flights. Moreover, tight budgets, tight schedules and a shortage of spares all combined to intensify problems. In 1981, a pipe in the combustion chamber broke. A year later, it broke again while being tested in *Challenger*. A spare engine was found to have the same fault. A second spare allowed *Challenger* to fly, but that was the last spare, and it had been intended for *Columbia*.

EATING ALOFT

Early astronauts had to make do with bite-sized cubes of food, freeze-dried powders, and tubes packed with semi-liquids. But it proved hard to rehydrate freeze-dried food, and cubes produced crumbs. On Apollo, astronauts were provided with hot water. Skylab had 72 different food items, kept fresh in a freezer and a refrigerator.

On the Shuttle, each crew member has three-quarters of a pound of food per day, in a carefully balanced diet, which begins to repeat only after a week. Astronauts can specify their own diet. On joint U.S.-Soviet missions, each side provides two meals a day. Research has shown that on long missions astronauts require less sodium and iron, and that Russian bread lasts longer and their fruit drinks are higher in fiber.

All food, except for fresh fruit and vegetables stowed in the food locker, is precooked. Much of it—like soups, casseroles, and scrambled eggs—is dehydrated. Since the orbiter's fuel cells produce electricity by combining hydrogen and oxygen, they also produce water for rehydrating food and for drinking. Other foods—beef, ham and grilled chicken—are heat-processed and packed in ring-pull cans. The food can be enriched with sauces, though salt and pepper have to be packaged in liquid form to prevent them from drifting around as dust. Snacks, such as nuts and biscuits, are also available.

With food ready in half an hour or less, and meal trays that lock the food in place, eating in space is now almost as easy as in a cafeteria.

LUNCH WITHOUT GRAVITY   Rhea Seddon, crew member on a 1985 Shuttle mission, shows how to eat in zero-G. Her feet are locked in straps to prevent her from floating about, and her food tray is secured to her thigh by velcro. Even the food on the tray is designed not to float free.

Columbia was grounded for three years, and when *Challenger* again needed a new engine, in February 1984, one had to be taken from *Discovery*. In August, as the 12th Shuttle mission—*Discovery*'s maiden flight—was four seconds from liftoff, a computer spotted that a fuel valve had not opened properly, and shut the engines down just before the boosters ignited.

Those in the government who used the Shuttle did not like what they saw. Defense Secretary Caspar Weinberger said that to rely exclusively on the Shuttle posed "an unacceptable national security risk." The Air Force said it planned to resume using Titan IIIs—now upgraded to rival the Shuttle's lifting power—ensuring a rosy future for the rocket's new owners, Martin Marietta, and even tighter competition for the Shuttle.

The ultimate nightmare

With the arrival of the fourth Shuttle, *Atlantis*, in late 1985, the fleet was complete. NASA looked toward 15 flights a year, rising to 24 a year by 1990. Given the constraints, technical problems and uncertain economic base, they needed to display all the confidence they could. In fact, the confidence was misplaced. Few missions had lifted off on time, and some parts were known to require further testing. But administrators were under constant pressure to establish a regular schedule, speed up turnaround times and dominate the markets, both national and

international, in science and commerce. Part of this effort persuaded NASA to offer flights to ordinary people. Teachers were invited to apply; of the 11,400 who did, Christa McAuliffe and her back-up, Barbara Morgan, were chosen.

The mission on which McAuliffe was to ride was planned for December 1985, but, as usual, the schedule slipped, with a launch date finally set for January 26, 1986. In Florida, bad weather struck, ushering in abnormal cold. The date was changed to the 28th. Still the cold weather persisted—subzero cold, almost unprecedented cold—with terrible consequences.

One of the persistent technical glitches concerned the join between the three sections of each of the solid-fuel boosters. These joins were secured by seals that were in effect giant rubber bands encircling the 12-foot rocket cases, acting as a flexible gasket. They were known as "O-rings," and their role was crucial: they had to withstand the sudden rise in pressure when the rockets ignited, flexing with the rockets' casing.

There had been trouble with the O-rings before. In April 1985, rocket exhausts had blown through one and eroded another. The rings had held, but among specialists at

A DOOMED TEAM Crew members of the ill-fated *Challenger* mission photographed a month before takeoff (above). Teacher Christa McAuliffe shares a joke over an apple with mission commander Dick Scobee. Her back-up, Barbara Morgan, who was not on the flight, is second from the left. Other members are Ellison Onizuka (far left), while to the right of McAuliffe are Judith Resnik, Michael Smith and Ronald McNair. As it lifted off (below), *Challenger* looked set for a perfect flight.

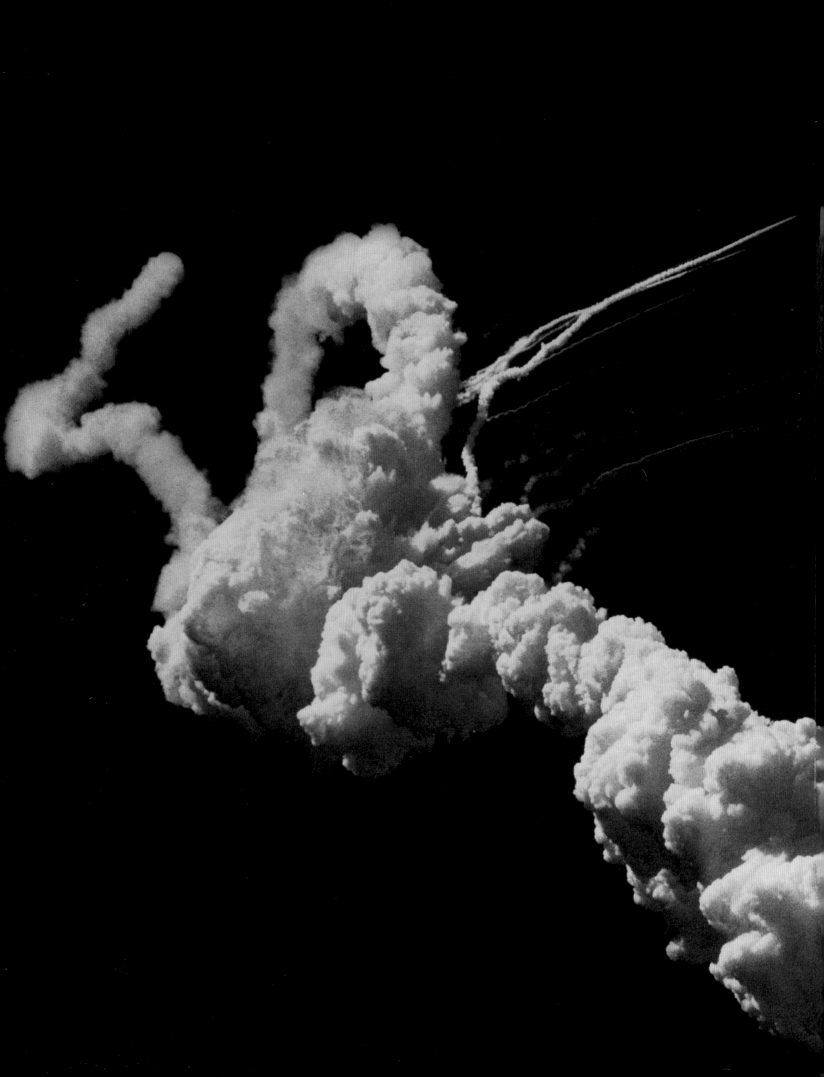

DISASTER AND GRIEF Seventy-three seconds into its mission, *Challenger*'s 1,000 tons of liquid fuel explode (left), blowing the rocket and its space vehicle to bits, and killing its seven occupants instantly. At a memorial service held in the Kennedy Space Center (above), President Reagan quoted from the Bible: "They slipped the surly bonds of earth and touched the face of God."

the booster manufacturer, Thiokol, there was concern. Three months later, Thiokol's senior seal specialist, Roger Boisjoly, wrote in a memo: "It is my honest and very real fear that if we do not take immediate action . . . we stand in jeopardy of losing a flight." Thiokol set up a team to look at the issue.

But there were countless such apparently little problems to be sorted out. There seemed no great urgency. Half a dozen flights succeeded each other through that summer, and no one saw any need for panic about the O-rings.

As *Challenger*'s winter launch drew nearer, however, Thiokol's concerns intensified. The company knew that the rubber rings worked best in summer temperatures. An analysis of past launches left them with the conclusion that below 53° Fahrenheit, the rings would perform less well. At the Cape, the night-time temperature was down to 29° Fahrenheit. Yet the rings had never actually failed, and everyone knew that the rockets could withstand a partial failure.

Thiokol faced a quandary. Technically, there was, in the jargon, no "temperature-failure correlation." In addition, NASA had recently invited other firms to compete for Thiokol's solid-booster contract. If Thiokol

pulled the plug on a Shuttle launch, thus declaring their product inferior, they risked going out of business. As launch time approached, through the night of January 27-8, crisis phone calls linked all those who were part of the decision-making process. On the launch tower, frost kept an icy grip, building 3-foot icicles. Engineers advised cancellation, or at least delay. "My God, Thiokol," said Lawrence Mulloy, who oversaw Thiokol's contract at the space center, "when do you want me to launch—next April?"

Faced with the grim choice between a risk that had proved acceptable before and certain commercial suicide, Thiokol's management opted to overrule their engineers. On the launch stand, preparations continued. None of *Challenger*'s occupants knew of the fears, the arguments or the closeness of the final decision.

The Sun came up, the icicles melted and concerns lifted. The engines fired, the Shuttle rose. But alongside the bottom joint of one of the rockets, a puff of smoke appeared, visible among the other billowing clouds only later, in freeze-frame and close-up analysis. A frozen O-ring had cracked, and a jet of fire was escaping through the

hole. Fifty-nine seconds into the flight, the ring gave further. The jet of flame became a blowtorch, burning at a support strut. After 12 seconds, the strut gave, and the booster swung against the main fuel tank, rupturing it. A thousand tons of liquid hydrogen exploded. *Challenger* disintegrated and was tossed away like a leaf.

Below, thousands saw the sight that seared millions watching on television: the distant speck of *Challenger* turning into a terrible firework, which fluttered down to the ocean in fragments. "I don't know how many seconds it took for the sound of the blast to travel down," wrote a reporter for the *Los Angeles Times*. "When it arrived, it did so like a thunderclap, rattling the metal grandstands. Then it abruptly ceased, replaced by a strange and terrible quiet."

In 120 ascents, neither the Americans nor the Soviets had lost a manned mission on launch. Now, in one second, as engineers stood aghast, as thousands stared in silence from the Cape's roadsides, the ever-possible nightmare had become a reality.

SHATTERED REMNANT One of the larger pieces of the wrecked *Challenger*, retrieved from the floor of the Atlantic Ocean, is brought back to the United States for analysis.

THE BUSINESS OF SPACE

RECOVERING FROM DISASTER, THE UNITED STATES FOUND ITSELF COMPETING INTERNATIONALLY TO BUILD A NEW INDUSTRY

The *Challenger* explosion in 1986 shook the nation to its core. The space industry began a bitter period of soul-searching and reappraisal. A presidential commission, headed by former Secretary of State William Rogers, pointed to managerial flaws within NASA as the fundamental cause of the disaster. As John Young, chief of NASA's astronauts said, there was only one reason for allowing *Challenger* to blast off in uncertain conditions: schedule pressure. It was this that persuaded NASA, and hence Thiokol, to consider the risk acceptable.

For the next two years, NASA reviewed the Shuttle's hardware and procedures. One consequence was that the Shuttle would not remain as a monopoly carrier. The Air Force had been given the task of developing the Strategic Defense Initiative, President Reagan's laser-based orbiting weapons system known as "Star Wars." For this project, it had built a $3 billion launch complex for the Shuttle at Vandenberg, California. Now the base went into mothballs, and expendable rockets were back. New Titans were ordered, and old ones—70 Titan II ICBMs—refurbished. In addition, Reagan ordered that the Shuttle could not carry commercial payloads that could be handled by Deltas, Atlases and Titans. This meant that the private operators of these rockets no longer had to worry about subsidized competition from the Shuttle. In 1988, one of the major operators, General Dynamics, committed itself to 62 new Atlas-Centaurs, which it said would dominate 90 percent of the commercial satellite market.

For a while, even expendable rockets had their problems. In April 1986, a Titan blew up just after liftoff, and the next month a NASA Delta failed. Successful launches in 1986-8 averaged under six a year. Ariane, the French rocket, also suffered a failure, its fourth in 18 launches. Ariane shut down for a year for modifications.

From *Buran* to Mir

In the same period, fortune smiled at last on the Soviet space program. The nation's long-serving leader in rocket-engine design, Valentin Glushko, had designed a new, heavy-lift launch vehicle suitable for the times. Korolev's successor, Vasily Mishin, had been superseded, and the giant SL-15 Moon rocket was consigned to the scrap heap. In its place was a major launch vehicle, Energiya. This formidable machine flew from Tyuratam in May 1987, carrying a new world-record payload of more than 100 tons.

SOVIET SHUTTLE The Soviet shuttle *Buran* ("Snowstorm") waits in its hangar (right). After a brief, unmanned flight in 1988, it made a safe, automatic landing (below).

Its spacecraft, a prototype battle station, failed to reach orbit; nevertheless, Energiya had proven itself.

Glushko's rocket had another purpose. It could act as carrier for the Soviet Union's own space shuttle, which had been given the name *Buran* ("Snowstorm"). *Buran* looked like the American Shuttle, but it had only small maneuvering engines; the major work of getting it into space was done by Energiya. In November 1988, the two flew successfully together and *Buran*, although unmanned, orbited the Earth twice before making a computer-guided landing.

Energiya and *Buran* were designed to mesh with the second strand of Soviet space

RUSSIA'S PLIGHT

One expert on Soviet space history, James Oberg, described the Baikonur Cosmodrome at Tyuratam, Kazakhstan, in 1995:

"There was plenty of decay and abandonment to see. By the early 1990s, there was often no heat or running water in workers' homes, no social services, from schools to medical care, and only the drabbest items of food in the stores. Security collapsed and squatters moved in while looters lurked in the city's outskirts. Public health declined rapidly and diseases spread, especially among the children. Many civilian workers, unpaid for months, abandoned their posts.

"The abandoned buildings, broken fences and thickly strewn junk piles reminded me of the worst extremes of U.S. urban decay. Abandoned apartments, sometimes in entire blocks, stare windowless at the dusty sun. The dust, blown from the pesticide-laden salt flats of the Aral Sea, is gradually poisoning the city's inhabitants, the weakest first.

"Somehow the hard-core space workers struggle on, enduring, improvising, cannibalizing, and making do. Members of this fanatic cadre, on average well over 50, have been through so much that no future challenge frightens them."

PROMISING START *Buran* waits on its Energiya rocket for the launch countdown at Tyuratam in Kazakhstan. The mission was a great success, but *Buran* never flew again, largely because of the collapsing Soviet economy.

activity, the creation of a space station. The last Salyut was still in orbit at the end of almost a decade of experience with space stations. Its successor, Mir, was launched in February 1986. Mir was a greatly improved version of Salyut, with more privacy and comfort. Still operating 13 years later, Mir's core consisted of four compartments, one each for docking, work, living, and engines, which are used for changing orbits—a vital function, for in its low orbit the station needs regular boosts to a higher level. The feature that made it most distinct was its ability to expand. It had six docking ports, allowing it to take on additional sections.

The first arrivals in March activated Mir and unloaded supplies sent up by unmanned Progress craft. The crewmen, Leonid Kizim and Vladimir Solovyev, then flew across to the surviving Salyut 7 (launched in 1982), where they stayed for seven weeks.

The next visitors to Mir, Yuri Romanenko and Alexander Laveikin, arrived in early

| 1990 Hubble Space Telescope launched by the Shuttle *Discovery* | 1991 Soviet Union starts to fall apart | 1992 Maiden flight of the Shuttle *Endeavor* | 1993 Crew of the Shuttle *Endeavor* capture and repair the Hubble Space Telescope | 1995 Valeri Polyakov returns from 14 months in orbit aboard Mir |

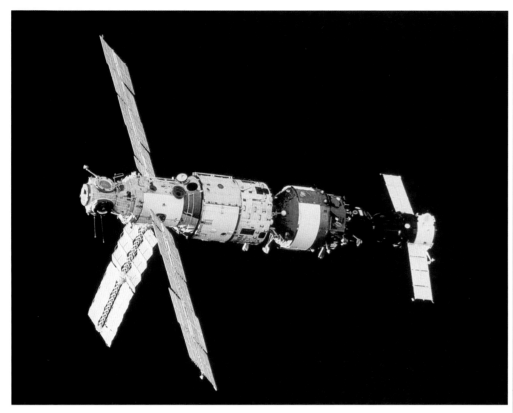

DRAGONFLY IN ORBIT With its solar panels spread like delicate wings, the core of the Soviet space station Mir originally had four compartments—for docking, working, living and for its engines.

1987 to prepare for the arrival of the first module, named Kvant ("Quantum"), which carried instruments to gather X-rays from a supernova, or exploding star. The appearance of this star was an event that astonished astronomers. First seen in February, it exploded in a neighboring galaxy, the Large Magellanic Cloud, 170,000 light years away. Though supernovae appear often in distant galaxies, this was the closest and brightest recorded in 400 years, and offered the first chance of studying one in detail. It turned out to have been a star 20 times the mass of the Sun and 50 times its size. Soviet scientists responded to the opportunity fast, and Kvant made a valuable contribution to the study of this unprecedented event.

The new crew also intended to make the first year-long stay in space. As it happened, Laveikin was replaced after developing health problems, but Romanenko set a new record by staying aloft

ENDURANCE RECORD During his record 326 days in Mir (almost 11 months), Yuri Romanenko undergoes a regular medical checkup. He returned exhausted and too weak to stand.

for 326 days. Replacement crews, however, did stay up for a year. Other research modules arrived, one of them with a dock designed to receive the *Buran* shuttle, although this never appeared. By the early 1990s, Mir had expanded to six sections.

In addition, the Soviet Union was at last in a position to cash in on the experience it had built up in space, for the very failings in its program had also produced strengths. For example, because they lacked reliable long-lasting electronics, the Soviets had always gone for quantity rather than quality. With 100 or so launches annually, they had developed a versatile fleet of workhorse rockets. In the late 1980s, the Soviet government set up a company, Glavkosmos, to market its launchers, but with little success. Within a few years, however, these launchers would look very attractive indeed.

The Soviet economic collapse in the early 1990s was catastrophic for the nation and for its entire space industry. Mir was virtually abandoned. A full-sized shuttle that had been used for tests could find no better use than as a display in Moscow's Gorky Park. In six years, military spending on space plummeted 90 percent. The civilian program dropped to $400 million a year in 1995, at a time when NASA was receiving $14 billion—35 times that amount. Space launches dropped from more than 100 to 23 in 1996. Tyuratam, the decaying space center in what had become an independent Kazakhstan,

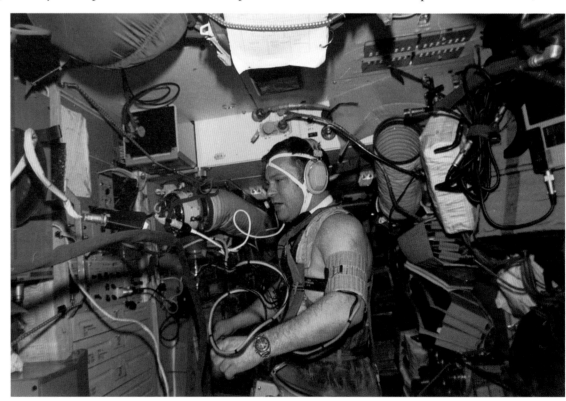

LAUNCH SITES AROUND THE WORLD

Each of the world's 15 launch sites is a social and technological world of its own: a construction site, a control center and a sophisticated technical center, all supporting a launch complex.

The choice of any site involves balancing several, often conflicting options. To position a launch site on the Equator saves money, because the Earth's spin gives a free 1,000 mph boost to the East, allowing more weight to be carried for less fuel. Yet few countries with space technology have access to such sites.

Launch sites should ideally be on an eastern coast to avoid damage in case of a disaster. But this can bring its own problems. Cape Canaveral, for example, is notorious for its ferocious hurricanes.

was the scene of a major riot in 1992, when the arrest of a military worker inspired several hundred soldiers to burn three barracks. Equipment was plundered, launch complexes abandoned. As one of the employees remarked: "I came here whole, healthy and unharmed and now after serving for two years I am going home sick, a cripple."

The collapse drove the space industry, like the whole country, into the arms of anyone who would pay in hard currency. Both the European Space Agency and NASA became paying customers.

Old enemies unite

By late 1988, the United States had begun to recover from the *Challenger* disaster. In September, the Shuttle *Discovery* made a flawless flight. The following year saw five Shuttle missions and 13 using expendables. NASA was back in the business of planetary exploration, sending the Magellan radar probe to Venus, followed by the delayed Galileo mission to Jupiter. Spacelab missions resumed in 1990.

The new missions returned with high hopes of paying the Shuttle's way through scientific advances, fulfilling a promise made even before the Shuttle ever flew. The

Shuttle would, NASA claimed, revolutionize molecular biology and produce, in President Reagan's words, "lifesaving medicines." On Earth, the business of understanding, let alone duplicating, the molecular structure of complex molecules was fraught with difficulty because gravity makes them unstable. For instance, it took the crystallographer Max Perutz 30 years to analyze hemoglobin's 10,000 atom structure and understand how it processed oxygen. If complex molecules could be grown in stable form in zero gravity as crystals, their structure could be analyzed and then perhaps manufactured artificially.

On the fourth Shuttle flight, crew members had experimented with a protein that stimulated the growth of red blood cells, the

lack of which causes a type of anemia. Earthbound attempts to grow the protein, erythropoietin, had shown that it broke down easily. But in zero gravity it thrived, with a given sample producing 400 times as much as on Earth and with five times the purity. In November 1983, on the ninth shuttle flight, the German specialist Ulf Merbold experimented with crystal growth, and grew one crystal to 27 times the size of anything achieved on Earth. Unfortunately, none of this research has yet resulted in a dramatic medical breakthrough.

Meanwhile, expendable rockets were back in force. By January 1989, France's Ariane had 38 launchers on order. U.S. Air Force orders included 11 Atlas-Centaurs, 20 Delta IIs, 13 Titan IIs and 23 Titan IVs. Then the collapse of Communism in 1991 led to the easing of export restrictions by the United States, allowing ex-Soviet launchers onto the world stage. Quite suddenly, it was a

UNINHIBITED GROWTH These germanium selenide crystals—semiconductors used in electronic devices—achieved record size and uniformity when grown in space, free from the deforming effects of gravity.

JUNK IN ORBIT

The pressure of space traffic has led to a problem utterly unforeseen by space pioneers: space debris.

From the earliest days of orbiting spacecraft, satellites posed a risk when they fell to Earth. In 1962, two policemen in the little town of Manitowoc on the shores of Lake Michigan found a 20-pound chunk of red-hot metal embedded in the street. It was the remains of Sputnik 4, which had burned up on re-entry. In January 1978, debris from Cosmos 954, including 100 pounds of uranium, scattered in the skies above Yellowknife in Canada's Northwest Territories. Since then, countless small satellites and several large ones (including Skylab) have broken up in the atmosphere.

Large satellites are now routinely re-boosted back into a safe orbit, but that contributes to the growing satellite population. Old satellites, pieces of rocket boosters, even lost clothing and tools, make up some 10,000 trackable objects and 150,000 bits of debris less than 4 inches across. Something even half this size, traveling at up to 10 miles per second, could knock out a satellite, or puncture a spacecraft and endanger lives.

Whenever the Space Shuttle journeys into orbit, it returns with 40 to 50 minor indentations caused by space debris. About half are from "natural causes," such as dust particles left over from passing comets. But the other half are caused by man-made items. It is a problem that can only get worse. Space debris experts foresee a time when some orbits are simply too cluttered for further use.

DANGEROUS DETRITUS There may be plenty of space between the thousands of man-made objects orbiting the Earth, but as they are traveling many times faster than a rifle bullet, they can be lethal. One tiny particle blasted this crater in one of the Hubble Telescope's solar panels (inset).

buyer's market. Around the world, in any one year in the early 1990s, any commercial satellite maker could pick and choose among 40 possible launches. There was even a newcomer in China's Great Wall launch company, which used its Long March rocket to re-launch an old communications satellite that had been rescued from space in 1984 by the Shuttle *Discovery*.

The collapse of the Soviet Union also intensified many changes that were sweeping the space industry. The United States and the Soviet Union agreed on arms cuts, and committed themselves to cooperating in space. Though seemingly a continuation of the Apollo-Soyuz venture, this was to involve more than a one-time project, for the two nations could reinvigorate each other's space programs on both government and commercial levels.

American companies scrambled for the former Soviet Union's treasure trove of rockets and engines. Khrunichev, the makers of the Proton rocket, had been the manufacturing overlords of a whole rocket sub-empire. Now the Soviet government couldn't pay for its orders, and the company was almost bankrupt. America's Lockheed stepped in, setting up a new launch company to use the Proton rocket. In April 1995, a Proton launched its first Western satellite. Aerojet bid for the old engines, made for the scrapped SL-15, which had been in storage for 20 years. Pratt & Whitney offered to market the RD-170, the world's most powerful engine, with 730 tons of thrust. Four of these engines had carried the abandoned Energiya into orbit.

These changes, combined with a flurry of corporate buy-outs in the United States, created a situation that would have been inconceivable a decade earlier. The end of the Cold War meant deep cuts in military contracts. Communications satellites now underpinned the economics of launches. Russia and America, the old adversaries, became colleagues. Atlas and Proton, devised to destroy each other, ended up in the same stable.

This opened a space race of an entirely different kind. The French, who only a few years before had been looking forward to dominating the launch market with Ariane, were appalled. The undercutters had themselves been undercut, first by the Chinese, and then by the Americans and Soviets act-

JOINT VENTURE At Tyuratam, a Soviet Proton booster carrying the Granat astrophysical observatory is raised onto its launch pad in December 1989. This project linked the Soviet Union, France, Denmark and Bulgaria. Meanwhile, China was preparing its own rocket, the Long March, first launched in July 1990 from the Xichang Space Center (left).

ing together. Charles Bigot, chairman of Arianespace, described what was happening as "a declaration of war." Unlike the Cold War, however, this war would not be fought in space, in the media or by Third World proxies; this was a war of multinationals, and it would be fought out in the marketplace.

The Hubble Space Telescope

In this fast-moving and complex world, NASA's old role as the guardian of large, publicly funded projects became unclear. A case in point was the Hubble Space Telescope, which was commissioned in the late 1970s. The roots of this magnificent instrument lay in telescopes mounted in spy satellites, which returned high-resolution images of the Earth to their military controllers. In contrast, the Hubble—named after Edwin Hubble, the astronomer who discovered that the Universe is expanding—would look the other way, into space. It was to be a true astronomical tool, with its own solar panels for power, on-board computer, and sufficient strength to be able to see planets orbiting around nearby stars.

Eager to streamline according to its budget, NASA's hard-pressed managers failed to subject Hubble to the tight quality controls it had insisted on in the past. The mirror's makers, Perkin-Elmer, had made many mirrors for the CIA, and had long experience checking their shape, which had to be accurate to within little more than a hair's width. Unfortunately, no one checked the accuracy of the instrument doing the checking.

The telescope was supposed to be ready in 1983. Delayed by budget cuts and various technical problems, the new launch date was to be January 1986, but the *Challenger* explosion at the beginning of that month delayed the Hubble launch yet again. When the telescope, costing $1.5 billion, was finally placed in orbit in 1990, it was found to have a highly embarrassing flaw: it was mildly nearsighted, the result of a microscopic deviation in its 94-inch mirror. Computers could rectify matters to some extent, but the error was a public relations disaster at a time when NASA was relying on good publicity as never before.

Three years later, in 1993, the crew of the Shuttle *Endeavor* captured the telescope and fitted it with additional mirrors that served as eyepieces, giving it perfect 20-20 vision. The result, at last, justified the hype. This was indeed a new eye on the universe, providing astonishing panoramas of galaxies, interstellar clouds and—most dramatic of all—the crash of the Shoemaker-Levy comet into Jupiter in July 1994. This was space technology at its best—scientific advances combined with superb publicity.

Pooling resources

International developments also brought an end to plans for an American space station. After some $9 billion had been spent, and redesigns had cut back the station's proposed costs to $30 billion, it was still no nearer to being built. Instead, a different proposal emerged: American cash would save the aging Mir by funding a new series of visits in both Soviet and American craft. And the two one-time enemies would combine as the prime contributors to a second-generation orbiting base, the International Space Station, which is due for completion in the first decade of the new millennium. Its price tag is around $100 billion, but the cost will be shared among 16 nations: the United States, Brazil, Russia, Canada, Japan and 11 European countries.

Meanwhile, Mir remained operational, if only barely. Both NASA and the European Space Agency became paying customers, with the German Skylab veteran Ulf Merbold spending a month on the station in 1994. Dr. Valeri Polyakov established a new

LEARNING TO COPE American astronaut Michael Foale here operates the video camera on Mir. In June 1997, he and two Russian cosmonauts successfully saved the station when it was punctured by an incoming supply ship.

space-endurance record by remaining in orbit for 14 months, from January 1994 to March 1995. As crew members came and went, they had to cope with increasingly serious "incidents," some 1,500 of which have been logged since Mir's launch.

A fire sent 12-foot flames shooting across the main cabin. A supply ship collided with one of the modules, Spektr. Computers and oxygen generators broke down. It was easy in 1997 for the media to portray Mir as a heap of junk. In fact, the glitches had their uses, for they forced the crew to develop unparalleled expertise in improvisation and space maintenance.

As the end of the millennium drew closer, Mir remained as a monument (albeit a slightly battered one) to the Soviet dream of a permanent presence in space. It was also a fitting inspiration for those planning its vast replacement, the International Space Station, which began to be built in 1998.

TO MARS, AND BEYOND?

WHILE 16 NATIONS COOPERATE TO BUILD THE BIGGEST SPACE STATION YET, SOME DREAMERS ARE ALREADY PLANNING TRIPS TO THE PLANETS

Visitors to Cape Canaveral can walk around a Saturn V rocket originally destined for one of the canceled Apollo missions. It lies on its side, like the fallen obelisk of a vanished civilization, which in a way, it is. The Moon missions were the product of an extraordinary combination of events in time—the perceived threat from the Soviet Union, a need to assert national prestige, a can-do mentality based on large government-backed projects, a confidence in science and technology, and a sudden startling vision expressed by a charismatic leader. Huge sums of money, growing expertise and a fascinated American public all focused on one aim. That sharpness of focus, that feeling that the very Universe was there for the grasping, lasted only a few years. By the time the Moon was reached, the world had changed, the moment had passed.

But other themes permeated the lunar mission, though they were unheard except by a few. The two main themes were human curiosity and a desire for financial profit. Satellites, which answer both urges, have wrought an economic and scientific revolu-

tion. Planetary probes and scientific satellites have provided some answers to deep questions about the nature of the Earth and human life on it, answers that give a new and more realistic sense of identity.

These themes, though minor ones in terms of the space race, have always influenced human expansion. In the great age of European exploration, from the 16th to the 19th century, lone explorers and small teams seeking fame, fortune and knowledge paved the way for settlers, scientists, businessmen and government. Large-scale settlements, in North America and Australia, found justification in economic realities. Small-scale ones, such as in Antarctica, allowed higher motives to flourish.

So it has been in space, with one major difference: in space, the true lone explorer and the small team were absent. The vital

role played by crews and individuals created an illusion. From the start, an astronaut could only be "alone" on the Moon because of the efforts of thousands. At any one time during his mission, an astronaut was able to communicate directly and indirectly with

WHY GO TO THE MOON?

The billions of dollars spent on the Moon missions and subsequent space ventures found justification in many practical benefits, or "spin-offs." Early spin-offs included computer bar codes, quartz timing crystals, non-stick materials such as Teflon, smoke detectors and cardiac pacemakers. As early as 1959, NASA had set up a special unit to coordinate the spin-offs, which now number over 30,000.

hundreds of people. Overwhelmingly, the point of manned space missions was not to take risks, but to bring astronauts back safely. Never has so much been spent on so few.

It is widely acknowledged that if things had been a little different, there would have been no Moon mission. There would have been satellites, certainly, and these would have led on to space stations of some kind. But the Moon? At that time? What for?

There are reasons, but the thinking involves longer-term aims and better justifications than those based simply on national self-interest. One argument goes back to

FALLEN GIANT A Saturn V rocket at Cape Canaveral acts as a memorial to an age when global politics inspired massive spending with little thought about economic returns.

ORBITING GIANT The International Space Station, due for completion early in the 21st century, could help humankind to become an interplanetary species.

President Kennedy's words in 1961. Space is still the "new frontier," not only for America but for all humankind. Indeed, as the television series *Star Trek* asserts, it is not only the new frontier, but the "final frontier." According to this argument, the human spirit demands new worlds to explore and tame. Discovery is a built-in human drive; the alternative is stagnation.

A second argument is economic. The Moon, like space stations, could provide bases, both for scientific study and for interplanetary travel. Unlike space stations, Moon bases might be made self-supporting, which would be a strong advantage. In either case, a space station or a Moon base would need to be cheap or it would need strong economic justification. It is this consideration that now underpins hard thinking about humankind's future in space.

What next?

In the 1970s, many dreamed of a bright future in space. There would be huge space stations, which would pay for themselves by beaming down power in the form of microwaves. Lunar colonies would turn a profit from mining operations, catapulting metals into lunar orbit for collection. With these, astronauts would build a huge space station, in which 10,000 people could live in a lush, tropical paradise, drawing power from the Sun and growing their own crops. Colonies would build colonies. Humans would move into the Solar System, then the Universe.

A generation later, such ideas seemed mere fantasy. Yet with a score of nations involved directly in space, with hundreds of satellites showing commercial advantage, with thousands of spin-off applications, with several promising avenues for technological and biological research, with staggering discoveries in astronomy—with all of this, there would, inevitably, be an expanding frontier. But how exactly is it likely to expand?

One suggested development involves combining both the rocket and the Shuttle concepts, using single-stage-to-orbit rockets as well as space planes, which would take off horizontally, fly into orbit and land again. At a time when Britain was involved in space only marginally, British engineers designed such a plane, known as HOTOL (Horizontal Takeoff and Landing). Of several possible designs, the most favored was a small plane riding piggyback on a larger one into the upper atmosphere, from where the small one would fly on into orbit.

But these refinements would do little to open any new frontier, other than an economic one benefiting the manufacturer of the space plane. The challenge of the real

COMPUTER DESIGN New craft, such as this British HOTOL (Horizontal Takeoff and Landing) space plane, could provide a safer and cheaper way of traveling into orbit.

frontier demands the conquest of other worlds in a way that can either be justified by low cost, or that promises significant, economic returns.

One low-cost method of exploration is by unmanned probe. Indeed, this is the only way of researching the outer reaches of the Solar System within a human lifetime. There are worlds enough out there for curiosity to feed on. Given the reliability and sophistication of deep-space gadgetry, there are already plans for other probes to Jupiter, Saturn, Neptune and Uranus, to their many moons, even to the tiny distant world of Pluto, with its enigmatic companion, Charon. At least three probes are destined to

MECHANICAL FRIENDS The robots C3P0 (left) and R2D2, from the *Star Wars* films, belong to a venerable sci-fi tradition of seeming almost human.

fly close to asteroids and comets in the new millennium, and one is even planned to land on a comet to analyze its make-up.

The Hubble Space Telescope can see much, and has revealed great wonders. But no distant observation matches the surprise of seeing a new world close up, even if only through the lens of a camera.

Yet space exploration must also, in the end, involve human beings directly. No machine can yet match the human mind in ingenuity and creativity. Despite popular anthropomorphic robots such as R2D2 and C3P0 in *Star Wars*, no machine will evoke the same empathy as Armstrong did when he first set foot on the Moon. Discovery acquires a sharper edge when accompanied by an eyewitness view and a hint of danger. It was this fundamental human emotion that Kennedy appealed to when announcing the Moon mission in 1961, and which President George Bush recalled in 1989, when he promised a manned mission to Mars.

Blueprint for a Mars mission

Mars has been the focus of exploration for a decade. As it happened, the assumptions current at the end of the 1980s about how a man or woman would actually reach Mars—giant space stations, giant ships—proved hopelessly expensive. But even as those hopes died, one leading American authority on space travel, Robert Zubrin, sketched an

THE FINAL FRONTIER This view of distant galaxies, taken by the Hubble Space Telescope, shows the immensity of space. Each one of these galaxies might be 100,000 light years across, yet the whole view is only $1/100$ of the diameter of the Moon.

idea for an affordable and safe way to get human beings to Mars, establish a colony, and make that colony pay its way.

After the Moon landing, it seemed that the foundations had been laid for a similar mission to Mars. But a Mars venture modeled on the Moon mission would involve prohibitive costs. One of the main problems was the need to carry fuel for the return journey. Another was the assumption that a crew member would have to be left in orbit, as in the lunar landings. Zubrin's plan, which he calls Mars Direct, avoids both approaches. His key aim was to use available techniques, travel cheap, and "live off the land"—in this case, the sky.

The rocket, which he calls Ares, would carry as much as the Saturn V, using Shuttle engines and boosters. Ares' first mission will be unmanned, its task being to take a small 45-ton spaceship to Mars. This vehicle will eventually return astronauts to Earth, and it will carry enough supplies to sustain four people on the eight-month return voyage. It will also have a few small robot rovers. Its key cargoes, however, will be an automated chemical unit, a small nuclear reactor and 6 tons of liquid hydrogen.

Once on Mars's rusty soil, the chemical plant, powered by the reactor, starts a simple series of reactions to manufacture rocket fuel. It draws in carbon dioxide, allows it to react with the hydrogen, and forms methane and water. The water is split into hydrogen and oxygen. The hydrogen is recycled, the oxygen stored. After six months, the plant will have produced more than 100 tons of methane and oxygen, which will be fuel enough to power its return journey.

Once space engineers know the operation is successful, the robot rovers will spend a few months surveying the ground to identify a suitable landing site. Some two years after

THE BIRTH OF STARS Photographed by the Hubble Space Telescope, this cloud of stellar matter, called a nebula, lies in a nearby galaxy, 2.7 million light years away. At its center, some 200 stars are being formed.

Ares 1, a second Ares will set off from Earth carrying the same payload. Soon afterward, Ares 3 takes off. This will be the first manned mission, with four crew members and living quarters, food for three years and a Mars rover. The four face a round trip of $2^{1}/_{2}$ years (less than Magellan's voyage around the world in 1519-22). As they proceed, their living capsule and upper-stage booster revolve twice a minute, creating Martian gravity by centrifugal force.

Ares 2 and 3 land on Mars at roughly the same time, but miles apart. Ares 3, with its crew, will be near Ares 1. Ares 2 creates a second landing site, but also provides for emergency supplies in the event of failure, allowing the crew to stay on Mars for up to three years, using their rover and the fuel provided by Ares 1 to explore an area several hundred miles across. Unlike astronauts on the Moon or in orbit, they will have Mars's atmosphere to protect them from solar radiation. They will prospect for ice, which is known to exist, or even water. They will experiment with growing food in an inflatable greenhouse. In the most intriguing work of all, they will search for signs of life.

If all goes well, after a year and a half the four will blast off for Earth, using the spacecraft landed by Ares 1. They will leave

A JOURNEY TO THE STARS?

In a sense, the age of interstellar travel has already begun. Four probes are on their way out of the Solar System, traveling at some 10 miles per second, or 36,000 mph. They also reveal the basic problem: even at their high speed, they would take 80,000 years to reach the nearest star, Alpha Centauri.

The interstellar gulf is too vast for current technologies. Alpha Centauri is four light years away. It would take eight years for us to receive a reply from a radio message to Centaurians, assuming they existed and had a suitable technology (they don't: we would have heard them). At four times the distance of Alpha Centauri, there are only 20 other stars. Suitable targets for travelers may be ten times the distance of Alpha Centauri, or 100, or 1,000, or even 25,000 times distant, for our galaxy is 100,000 light years across.

Space engineers have wrestled with ways to bridge these vast gulfs, and have come up with only a few suggestions. One is to use a "light-drive" in which light itself would accelerate a spaceship to a speed close to that of light—186,300 miles per second. A major problem with this idea is that, according to relativity theory, the spaceship's mass would rise to infinity as it approached light speed. And even if such speeds could be achieved, journeys would still last for centuries. Another equally unproven idea is that of a vast magnetic scoop gathering hydrogen from the interstellar medium to create its own fuel.

The least impractical solution is an unmanned ship, code-named Daedalus, devised by the British Interplanetary Society in the 1970s. Daedalus would use nuclear fusion to provide a rapid series of small nuclear pulses, around 250 a second. To get to a nearby star would take 30 billion nuclear "bombs" the size of ping pong balls, which would drive a ship weighing more than 50,000 tons on departure from orbit. It would accelerate to 12 percent of light speed, and take 50 years to cover six light years.

Even if Daedalus ever became practical, current thinking about space suggests that no nation would fund such a project, unless contact with some other civilization had already been established. And if that happened, the civilization might well be hundreds of light years away. Information about its existence would have taken centuries to reach us. There would be no guarantee that the civilization would still exist by the time our probe arrived there; nor that we would still exist to receive any message sent back by the probe. Such considerations have kept interstellar travel firmly in the realm of science fiction.

TO THE STARS In this artist's impression, the Daedalus starship, powered by its flickering glow of rapid nuclear pulses, heads past Neptune on its journey out of the Solar System.

behind a working base, a stock of fuel, growing plants and most of their scientific gear. Meanwhile, a second crew will have left Earth to perform a similar role at the second base established by the automatic Ares 2.

This is a process that can continue indefinitely, with two rockets taking off every two years, one with a crew and living quarters, another with an Earth Return Vehicle. This, according to Zubrin, is an affordable, sustainable program of Mars exploration. It has the advantages of safety and flexibility, and could certainly be undertaken by the United States alone, if the will is there. It could also be undertaken by an international consortium, or by European nations.

Over time, one of the bases will emerge as the best for settlement. Astronauts will be able to lengthen their stay, combining and extending their living quarters into a small township. In the long term, this settlement could act as a basis for something that, though still far in the future, is considered feasible by many: with plants releasing oxygen, Mars's atmosphere could slowly be transformed. Over centuries, it could again become the living, breathing world it once was, or might have been.

Zubrin's scenario is plausible. But what would be the purpose? Scientific research could well be a good enough answer, given the low costs. There could also be an economic appeal, for Mars is rich in deuterium, which is used in nuclear reactors and which costs about $10,000 an ounce on Earth.

There could also be subtler and deeper arguments for such a project. A Martian colony would be creating a new world. As it grew, it would produce writers, artists and filmmakers. It would stimulate new social thought. Who would own Mars? How would it be administered? For an Earth-bound world of increasing uniformity, suffering from the effects of limited resources and pollution, a Martian colony could inspire afresh, as the discovery of the New World did on Earth. From new necessities, new inventions could flow. As the space race spawned whole new industries, Martian technology could reinvigorate a weary Earth.

Finally, the Mars missions would open a window on the Universe. The technique of the missions could apply also to the Moon (though the Moon's harsher conditions would probably attract scientists, not long-term settlers). With settlers on Mars, other planets, moons and asteroids would beckon, with unguessable consequences.

That, in the end, may be the conclusion toward which the visions and fears of half a century of space travel are leading—a race away from national competition, and toward the new, and final, frontier.

TIMECHART

PRE-1900

1865: Start of serialization of Jules Verne's *From the Earth to the Moon* and *A Trip Around the Moon*.

1898: H.G. Wells publishes *The War of the Worlds*.

1903

Konstantin Tsiolkovsky publishes *Exploring Cosmic Space*.

1904

In the United States, Robert Goddard starts rocket experiments.

1914

The First World War breaks out.

1918

End of the First World War.

NEW AGE In 1937, a French magazine proclaims the arrival of the space age by announcing that a journey to the Moon might be possible.

1919

Goddard publishes *A Method of Reaching Extreme Altitudes*.

1922

In Germany, Hermann Oberth reads Goddard's paper and realizes that his own work is more ambitious.

1923

Hermann Oberth publishes *The Rocket into Interplanetary Space*.

1926

Goddard engineers the world's first liquid-fuel rocket flight.

1928

Willy Ley publishes *The Possibility of Space Travel*.

1929

Release of Fritz Lang's film *Frau im Mond* (*Woman in the Moon*).

1931

Wernher von Braun and other members of Society for Space Travel fire their first rocket.

1935

Tsiolkovsky dies.

1942

Code-named A-4, the first V-2 rocket is fired in Peenemünde in Germany.

1943

Allied air raid destroys V-2 site at Peenemünde.

1944

V-2 assault opens on London.

1945

American troops seize V-2 production in Germany.

War ends in Europe.

100 V-2s shipped to the United States.

First atomic bombs dropped on Japanese cities.

Goddard dies.

Arthur C. Clarke publishes description of satellites in geostationary orbit.

1946

V-2 carrying Corporal "sounding rocket" fired in White Sands, New Mexico.

1952

Collier's magazine starts a series of articles on humanity's future in space.

U.S. tests H-bomb in the Pacific.

1953

U.S.S.R. tests H-bomb.

1954

America's Viking 11 takes first photographs of the Earth from space.

1957

U.S.S.R. launches first artificial satellite, Sputnik 1.

U.S.S.R. launches Sputnik 2, carrying a dog, Laika.

1958

American rocket Explorer 1 discovers Van Allen Radiation Belts.

National Aeronautics and Space Administration (NASA) founded.

Pioneer 1 (U.S.) travels 70,000 miles into space and returns.

1959

Luna 1 (U.S.S.R.), fired at the Moon, misses its target and becomes the first man-made object to go into orbit around the Sun.

TIROS 1 (U.S.) becomes the first weather satellite.

Luna 2 (U.S.S.R.) is the first man-made object to strike the Moon.

SOVIET HEROES Yuri Gagarin and Gherman Titov, the first two Soviet cosmonauts in orbit, chat quietly.

JUBILANT SCIENTISTS Wernher von Braun (right) and James Van Allen (center) raise Explorer 1 in triumph. Background: Konstantin Tsiolkovsky, the father of modern rocketry, with two models.

Luna 3 (U.S.S.R.) sends back the first images of the Moon's far side.

United Nations Committee on the Peaceful Uses of Outer Space set up.

1960

Echo (U.S.) proves communications satellites can work.

Soviet spacecraft returns two dogs alive from space.

1961

U.S.S.R.'s Venera 1 launched to Venus; contact is lost.

Soviet cosmonaut Yuri Gagarin becomes the first human in space when he orbits the Earth in Vostok 1.

Alan Shepard becomes the first American in space.

President Kennedy dedicates the country to putting a man on the Moon "before the decade is out."

In Vostok 2, U.S.S.R.'s Gherman Titov orbits the Earth 17 times.

1962

John Glenn becomes the first American to orbit the Earth.

Ariel probe (Britain-U.S.) launched to study cosmic rays.

Telstar (U.S.) becomes the first communications satellite.

Unmanned American probe Mariner 2 launched toward Venus.

Mars 1 (U.S.S.R.) launched; contact lost.

Mariner 2 flies past Venus and sends back the first detailed information on the planet's atmosphere.

APOLLO RECOVERY Rescuers, dropped from a helicopter, lend a hand to the Apollo 8 crew after splashdown in the Pacific Ocean in December 1968.

1963

U.S.S.R.'s Valentina Tereshkova becomes first woman to orbit Earth.

U.N. declares principles that should govern exploration and use of space for all nations.

1964

Mariner 4 (U.S.) launched toward Mars.

1965

Alexei Leonov (U.S.S.R.) becomes first to walk in space.

Gemini 3: First U.S. Gemini mission (Grissom and Young).

Early Bird (U.S.) becomes the first commercial communications satellite.

American Edward White takes a 20-minute space walk.

First pictures of Mars received from Mariner 4.

In Gemini 5 (U.S.), Gordon Cooper and Charles Conrad spend eight days in space, making 120 orbits.

Venera 3 (U.S.S.R.) launched to Venus.

First French satellite launched.

1966

Sergei Korolev, mastermind of the Soviet space program, dies.

Soviet Luna 9 makes the first unmanned landing on the Moon.

United States launches the first weather-reporting satellite.

Venera 3 (U.S.S.R.) crashes on Venus.

Gemini 8 docks successfully with an unmanned craft.

Soviet Luna 10 becomes the first spacecraft to orbit the Moon.

Surveyor 1 (U.S.) soft-lands on Moon.

Venera 4 (U.S.S.R.) launched to Venus.

1967

Apollo 1 crew—Virgil "Gus" Grissom, Roger Chaffee and Edward White— die in a fire during training.

Signing of U.N.'s treaty on exploration and use of space.

Soyuz 1 (U.S.S.R.) ends with death of Vladimir Komarov.

Mariner 5 launched toward Venus.

Venera 4 makes the first soft-landing on Venus.

Mariner 5 passes Venus and sends back pictures.

Two unmanned Soviet spacecraft dock in orbit.

1968

First manned Apollo mission (Apollo 7) launched from Cape Canaveral.

Apollo 7 becomes the first manned spacecraft to orbit the Moon.

1969

Veneras 5 and 6 (U.S.S.R.) take off for Venus.

MISSION BADGE Apollo 10's emblem shows the ascent stage of the lunar lander rising from the Moon.

Soyuz 5 and 6 (U.S.S.R.) rendezvous in space.

Mariner 6 takes off for Mars.

Apollo 9 docks in space with lunar landing module.

Veneras 5 and 6 land on Venus, transmitting during descent.

Apollo 10's lunar lander, *Snoopy*, descends to within 9 miles of the lunar surface.

Apollo 11 takes off for the Moon with Neil Armstrong, Edwin "Buzz" Aldrin and Michael Collins on board. While Collins remains in orbit, Armstrong and Aldrin become the first humans to set foot on the Moon.

Mariners 6 and 7 fly by Mars.

Apollo 12 lands on the Moon.

1970

China launches its first satellite.

An explosion cripples Apollo 13 on its way to the Moon. Improvisations by NASA engineers and the astronauts enable the crew to return safely to Earth.

Venera 7 probe launched to Venus.

Unmanned Soviet probe, Luna 16, soft-lands on the Moon and returns a sample of soil to Earth.

Unmanned Soviet Luna 17 lands Lunokhod, first vehicle on Moon.

Venera 7 lands on Venus.

1971

Apollo 15 lands on the Moon with the Lunar Rover.

Mariner 9 goes into Mars orbit.

Mars 2 (U.S.S.R.) lands on Mars.

First Soviet space station, Salyut 1, goes into orbit.

1972

Venera 8 lands on Venus and transmits messages for 50 minutes.

Apollo 17 mission is last in the Apollo series of lunar missions.

1973

Mariner 10 launched to Mercury.

Pioneer 10 flies by Jupiter.

Skylab (U.S.) damaged on launch; it is repaired by its first crew.

1974

Mariner 10 makes the first of three passes of Mercury.

Pioneer 11 flies by Jupiter and continues to Saturn.

1975

Apollo-Soyuz (joint U.S.-Soviet mission) launched.

HIDDEN PLANET The surface of Venus, obscured by swirling clouds of gas, is here revealed by the radar of the Magellan probe (U.S.).

Veneras 9 and 10 (U.S.S.R.) land on Venus and send back the first images.

1976

Vikings 1 and 2 (U.S.) soft-land on Mars and send back images and analyses of Martian soil.

1978

Pioneer 1 (U.S.) starts to radar-map Venus from orbit.

Pioneer 2 delivers four probes to the surface of Venus.

1979

Voyager 1 flies by Jupiter and continues to Saturn.

Voyager 2 flies by Jupiter, before continuing to Saturn, Uranus and Neptune.

1981

Space Shuttle *Columbia* makes its first operational flight.

1982

Veneras 13 and 14 soft-land on Venus, transmitting color pictures to Earth.

1983

Veneras 15 and 16 start to radar-map Venus from orbit.

European Spacelab placed in orbit by the Space Shuttle *Columbia*.

1986

Space Shuttle *Challenger* blows up, killing all seven crew and halting the shuttle program.

Launch of Soviet space station Mir.

Europe's Giotto probe passes within 370 miles of Halley's comet.

1988

Space Shuttle *Discovery* successfully launched to restart Shuttle program.

1990

Japan launches lunar orbiter.

Space Shuttle *Discovery* launches Hubble Space Telescope.

Magellan probe, launched by Space Shuttle, starts to radar-map Venus.

1992

Maiden flight of Space Shuttle *Endeavor*. Longest space walk (8 hours 29 minutes). Intelsat captured.

1993

Crew of the Shuttle *Endeavor* repair the Hubble Space Telescope.

TESTS ON MIR Valeri Polyakov and the German Ulf Merbold (right) conduct medical tests on Mir. Background: the Space Shuttle *Columbia* lifts off at the Kennedy Space Center in Florida in November 1983.

1994

Sergei Kirkalev becomes first Russian to fly on an American space mission.

The Space Shuttle *Endeavor* places the Space Radar Laboratory in orbit.

1995

Valeri Polyakov leaves Mir after establishing a new space-endurance record—438 days in orbit.

Space Shuttle *Atlantis* adds docking module to Mir.

1996

First U.S. woman astronaut to work on Mir.

Space Shuttle *Columbia* concludes the longest-ever Shuttle mission (18 days).

1997

Pathfinder (U.S.) soft-lands on Mars. Its rover unit, called Sojourner, analyzes Martian rocks. Contact with Sojourner is maintained for three months.

Galileo probe (U.S.) passes Jupiter's moon Europa at a distance of 124 miles, starting a series of missions to the planet's other moons.

1998

John Glenn returns to space aboard the Shuttle *Discovery*, 36 years after becoming the first American in orbit.

First modules of the International Space Station are launched, the first from Kazakhstan, the second from Cape Canaveral.

VETERAN PERFORMER At the age of 77, John Glenn became the oldest man in space. He performed a battery of tests to study the effects of zero gravity on someone of his age.

ACKNOWLEDGMENTS

Abbreviations
T = top; B = bottom; R = right;
L = left; M = middle

3 Bilderdienst Suddeutscher Verlag, L; NASA, ML, R; Genesis Space Photo Library, MR. 6 Novosti, TR; © CORBIS, BL. 7 NASA/Science Photo Library, TR; Novosti, BL. 8 Popperfoto, TL; NASA, BR. 9 NASA, T; Genesis Space Photo Library, MR. 10 NASA, TL; Rex Features Ltd, TR; Genesis Space Photo Library, BR. 11 Novosti, background, L, R; NASA, ML; Mary Evans Picture Library, MR. 12 Mary Evans Picture Library. 13 Novosti. 14 UPI/Corbis. 15 NASA/Science Photo Library. 16 NASA/Marshall Space Flight Center. 17 Corbis-Bettmann, BL; Bilderdienst Suddeutscher Verlag, MR. 18 Bilderdienst Suddeutscher Verlag. 19 Illustration by Graham White. 20 Bilderdienst Suddeutscher Verlag, BL; Novosti, TR. 21 Bilderdienst Suddeutscher Verlag, TL; NASA/Marshall Space Flight Center BM; NASA, BR. 23 Ullstein Bilderdienst. 24 Ullstein Bilderdienst. 25 Popperfoto. 26 Illustration by Graham White, L; Popperfoto, BR. 27 Corbis-Bettmann/UPI. 28 NASA/Marshall Space Flight Center, TL; Popperfoto, BR. 29 Popperfoto, background, R; Novosti, L, MR; Mary Evans Picture Library, ML. 30 Popperfoto. 31 Popperfoto. 32 Popperfoto. 33 Corbis. 34 John Frost Historical News Archives, MM; Mary Evans Picture Library, BL, BML; Kobal Collection, BMR, BR. 35 Associated Press/Topham. 36 Topham Picturepoint. 37 Corbis-Bettmann/UPI. 38 Associated Press/Topham. 39 Novosti. 40 Popperfoto, BL; Ullstein Bilderdienst, TR. 41 Bilderdienst Suddeutscher Verlag, BL; Ullstein Bilderdienst, R. 42 NASA/Marshall Space Flight Center, ML; Topham Picturepoint, BR. 43 Topham Picturepoint, TL; Novosti/Science Photo Library, ML. 44 NASA/JPL, L; Associated Press/Topham, TR; Illustration by Graham White, MR. 45 Popperfoto. 46 Corbis-Bettmann, TL; NASA/Marshall Space Flight Center, BR. 47 NASA. 48 Science Photo Library, BL; Novosti, TR. 49 Topham Picturepoint. 50 Bilderdienst Suddeutscher Verlag, TL; NASA/Science Photo Library, BL. 51 Novosti. 52 Novosti, TR; Popperfoto, MB. 53 Illustration by Graham White. 54 NASA. 55 NASA. 56 Illustration by Graham White, L; UPI/Corbis, BR. 57 NASA, BL; Corbis-Bettmann/UPI, MR. 58 NASA. 59 Corbis-Bettmann/UPI. 60 Corbis-Bettmann/UPI, TR; Popperfoto, BL. 61 Novosti, TL; Novosti Press Agency/Science Photo

Library, BR. 62 Toucan Books Archives, TM; Corbis-Bettmann/UPI, BL. 63 NASA, TR; Bilderdienst Suddeutscher Verlag, MB. 64 Popperfoto, BR; NASA, RM. 65 NASA, TL, BR; NASA/Science Photo Library, MB. 66 Associated Press/Topham, TL; NASA, MR, B. 67 NASA. 68 NASA/Marshall Space Flight Center. 69 NASA/Marsahll Space Flight Center. 70 Hallman Johnson/NASA/Marshall Space Flight Center. 71 NASA, TL; Illustration by Graham White, M. 72 Novosti, TL; Genesis Space Photo Library, MM, BL, BM. 73 Hulton-Getty, BL; Popperfoto, BR. 74 NASA/Science Photo Library, TR; NASA, MM; Genesis Space Photo Library, BL. 75 Corbis-Bettmann/UPI, MR; Magnum Photos, B. 76 NASA. 77 NASA. 78 NASA. 79 NASA, BL; NASA/Marshall Space Flight Center, TR. 80 NASA. 81 Illustration by Kevin Jones Associates. 82 Novosti, TL; Illustration by Graham White, B. 83 NASA, TL; NASA/Marshall Space Flight Center, BR. 84 NASA. 85 NASA. 86 NASA. 87 NASA. 88 NASA. 89 NASA. 90 NASA. 91 NASA. 92 NASA/Marshall Space Flight Center, TL; NASA, B. 93 NASA/Marshall Space Flight Center. 94 NASA. 95 NASA. 96 NASA. 96-97 Genesis Space Photo Library. 98 Novosti/Science Photo Library, TR; NASA, BL. 99 NASA/JPL, background, MM; National Organisation for Atmospheric Research/Science Photo Library, LM; Ames Research Center, RM. 100 Corbis-Bettmann/UPI, BL; Illustration by Graham White, TR. 101 Illustration by Graham White, TL; Novosti, BL. 102 CNES, 1988 Distribution Spot Image/Science Photo Library, BL; US Geological Survey/Science Photo Library, M; CNES, 1987 Distribution Spot Image/Science Photo Library, TR. 103 NASA GSFC/Science Photo Library, TR; NASA/Science Photo Library, BM. 104 NRSC Ltd/Science Photo Library. 104-5 Genesis Space Photo Library. 105 National Organisation for Atmospheric Research/Science Photo Library. 106 AKG/ESA, TL; Novosti, R. 107 Novosti, TR; NASA/JPL, BL. 108 NASA/JPL, TL; Novosti/Science Photo Library, BL. 109 NASA/JPL, T; Illustration by Kevin Jones Associates, BR. 110 NASA/JPL. 111 NASA/JPL, TR; NASA/Science Photo Library, B. 112 NASA/JPL. 113 Illustration by Graham White, BL; Ames Research Center, MR. 114 NASA/JPL. 115 Ames Research Center, TR; NASA/JPL, ML. 116 Illustration by Graham White, T; NASA/JPL, BR. 117 NASA/JPL, TL; Illustration by Kevin Jones Associate, R. 118 Corbis-Bettmann/UPI, TR; European Space Agency/Science Photo Library, ML;

Illustration by Graham White, MR. 119 NASA, background, L, MR, R; Dryden/NASA, ML. 120 Novosti. 121 Novosti. 122 NASA. 123 NASA. 124 Rex Features Ltd. 124-5 Corbis-Bettmann/UPI. 126 NASA, TL; Novosti, BR. 127 Novosti. 128 Novosti. 129 Dryden/NASA. 130 Genesis Space Photo Library. 131 NASA. 132 NASA. 133 NASA, TL; Dryden/NASA, ML, MR, BR. 134 NASA. 135 NASA, TR, BR; ESA/NASA, MR. 136 NASA. 137 NASA. 138 NASA. 139 NASA. 140 Novosti. 141 Novosti. 142 Novosti. 143 Illustration by Colin Woodman, TR; NASA, BM. 144 Dr Seth Shostak/Science Photo Library, TL; Geoff Tompkinson/Science Photo Library, MM. 145 Novosti, T; China Great Wall Industry Corporation/Science Photo Library, BL. 146 NASA/Science Photo Library. 147 Rex Features Ltd. 148-9 NASA. 149 Tony Craddock/Science Photo Library. 150 Kobal Collection,TL; Space Telescope Science Institute/NASA/Science Photo Library, BR. 151 Space Telescope Science Institute/NASA/Science Photo Library. 152 David A. Hardy/Science Photo Library. 153 Novosti, background; Mary Evans Picture Library, ML; NASA/Marshall Space Flight Center, TL; Novosti Press Agency/Science Photo Library, BL. 154 NASA. 154-5 NASA, background. 155 Genesis Space Photo Library, TM; NASA/JPL, BL, BR.

Front Cover:
Top: NASA
Middle: NASA/JPL
Bottom: Genesis Space Photo Library

Back Cover:
Top: NASA
Middle: NASA
Bottom: Corbis-Bettman/UPI

The editors are grateful to the following individuals and publishers for their kind permission to quote passages from the publications listed below:

Farrar, Straus, Giroux from *Carrying the Fire* by Michael Collins, 1974. Harrap Publishers Ltd from *Red Star in Orbit* by James Oberg, 1981. Penguin Books Ltd from *A Man on the Moon* by Andrew Chaikin, 1994. Virgin Publishing Ltd from *Moonshot:The Inside Story of America's Race to the Moon* by Alan Shepard and Deke Slayton, 1994. John Wiley from *Countdown: A history of Space Flight* by T.A. Heppenheimer, 1997. *Wireless World*, October 1945.